Introduction to
differential topology

Introduction to differential topology

TH. BRÖCKER AND K. JÄNICH

Professors of Mathematics, University of Regensburg

TRANSLATED BY C. B. AND M. J. THOMAS

CAMBRIDGE UNIVERSITY PRESS

CAMBRIDGE
LONDON   NEW YORK   NEW ROCHELLE
MELBOURNE   SYDNEY

CAMBRIDGE UNIVERSITY PRESS
Cambridge, New York, Melbourne, Madrid, Cape Town, Singapore, São Paulo

Cambridge University Press
The Edinburgh Building, Cambridge CB2 8RU, UK

Published in the United States of America by Cambridge University Press, New York

www.cambridge.org
Information on this title: www.cambridge.org/9780521241359

Originally published in German as *Einführung in die Differentialtopologie* by
Springer-Verlag, Berlin, 1973 and © Springer-Verlag Berlin Heidelberg 1973.
First published in English by Cambridge University Press 1982 as
Introduction to differential topology
English edition © Cambridge University Press 1982

Re-issued in this digitally printed version 2007

A catalogue record for this publication is available from the British Library

Library of Congress Catalogue Card Number: 81-21591

ISBN 978-0-521-24135-9 hardback
ISBN 978-0-521-28470-7 paperback

Contents

Preface

The aim of this book is to describe the basic geometric methods of differential topology. It is intended for students with a basic knowledge of analysis and general topology.

We prove embedding, isotopy and transversality theorems, and discuss, as important techniques, Sard's theorem, partitions of unity, dynamical systems, and (following the example of Serge Lang) sprays. We also consider connected sums, tubular neighbourhoods, collars and the glueing together of manifolds with boundary along the boundary.

We have ourselves learned much from the writings of Milnor, as has nearly every young topologist today, and there are traces of this in the text. We have also from time to time drawn on Serge Lang's exemplary exposition [3] – to studiously avoid doing this would certainly not benefit any book about differential topology.

The numerous exercises at the end of each chapter are not always easy for a beginner; they are not used in the text.

We do not discuss analysis on manifolds (Stokes' theorem), Morse theory, the algebraic topology of manifolds or bordism theory. However, we hope that our book will provide a solid basis for a closer acquaintance with these more advanced topics of differential topology.

Regensburg, Pentecost 1973   Theodor Bröcker, Klaus Jänich

1
Manifolds and differentiable structures

A manifold is a topological space which 'locally resembles' $\mathbb{R}^n$, the Euclidean space of real n-tuples $x = (x_1, \ldots, x_n)$ with the usual topology. Such spaces result in general, as we shall later see, as solution spaces of non-linear systems of equations, and many of the concepts of general topology have developed out of the study of these special spaces. The precise explanation is as follows:

(1.1) **Definition.** An *n-dimensional topological manifold M^n* is a Hausdorff topological space with a countable basis for the topology, which is locally homeomorphic to $\mathbb{R}^n$. The last condition means that, for each point $p \in M$, there exists an open neighbourhood U of p and a homeomorphism

$$h: U \to U'$$

onto an open set $U' \subset \mathbb{R}^n$ (Fig. 1).

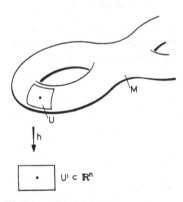

Fig. 1

The requirement that the space must be Hausdorff does not follow from the local condition as one might believe. As a counterexample one takes the real line $\mathbb{R}$, together with an additional point p, see Fig. 2, and defines the topology on $M = \mathbb{R} \cup \{p\}$ by saying that $\mathbb{R}$ is open and that the neighbourhoods of p are the sets $(U - \{0\}) \cup \{p\}$, where U is a neighbourhood of $0 \in \mathbb{R}$. Examples of topological manifolds (see Fig. 3) are:

1

Fig. 2

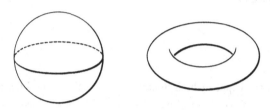

Fig. 3

every open subset of a Euclidean space;

the n-sphere $S^n = \{x \in \mathbb{R}^{n+1} \mid |x| = 1\}$;

the torus or surface of an inner tube (1.8).

(1.2) Definition. If M^n is a topological manifold and $h: U \to U'$ a homeomorphism of an open subset $U \subset M$ onto the open subset $U' \subset \mathbb{R}^n$, then h is a *chart* of M and U is the associated *chart domain*. A collection of charts $\{h_\alpha \mid \alpha \in A\}$ with domains U_α is called an *atlas* for M if $\cup_{\alpha \in A} U_\alpha = M$.

Given two charts, both homeomorphisms h_α, h_β are defined on the intersection of their domains $U_{\alpha\beta} := U_\alpha \cap U_\beta$ and one thereby obtains the chart transformation $h_{\alpha\beta}$ as a homeomorphism between open subsets of $\mathbb{R}^n$ by means of the commutative diagram:

$$U_{\alpha\beta}$$

$$h_\alpha \swarrow \qquad \searrow h_\beta$$

$$U'_\alpha \supset h_\alpha(U_{\alpha\beta}) \xrightarrow[h_{\alpha\beta}]{} h_\beta(U_{\alpha\beta}) \subset U'_\beta$$

in which $h_{\alpha\beta}$ is defined as $h_\beta \circ h_\alpha^{-1}$, see Fig. 4.

Occasionally, we shall find it useful to include the domain of definition of a map, particularly of a chart, in the notation, and thus we shall write (h, U) for a map $h: U \to U'$. If one were to consider the whole manifold as being formed by a glueing process from the chart domains, which one knows as well as one knows the open subsets of Euclidean space, then it is precisely the chart transformations which show how different chart domains are to be glued together. If, apart from the topological, one wishes to extend additional properties from open subsets of Euclidean space to manifolds by means of a

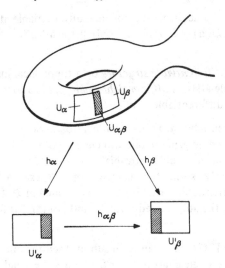

Fig. 4

suitable atlas, one must ensure that the definitions are independent of the particular choice of charts in the atlas, or that the property under consideration is independent of the chart transformations.

(1.3) **Definition.** An atlas of a manifold is called *differentiable*, if all its chart transformations are differentiable.

We shall always consider a differentiable mapping between open subsets of $\mathbb{R}^n$ to be a C^∞-mapping, that is, a mapping whose various (higher) partial derivatives exist and are continuous. Because, for the chart transformations $h_{\alpha\beta}$ (wherever the respective maps are defined), it is clear that

$$h_{\alpha\alpha} = \mathrm{Id}, \quad h_{\beta\gamma} \circ h_{\alpha\beta} = h_{\alpha\gamma},$$

it follows that

$$h_{\alpha\beta}^{-1} = h_{\beta\alpha}.$$

Therefore, the inverses of the chart transformations are also differentiable, and the chart transformations are diffeomorphisms.

If $\mathfrak{A}$ is a differentiable atlas on the manifold M, then the atlas $\mathfrak{D} = \mathfrak{D}(\mathfrak{A})$ contains precisely those charts for which every chart transformation with a chart from $\mathfrak{A}$ is differentiable. The atlas $\mathfrak{D}$ is then differentiable as well, since one can locally write a chart transformation $h_{\beta\gamma}$ in $\mathfrak{D}$ as a composition $h_{\beta\gamma} = h_{\alpha\gamma} \circ h_{\beta\alpha}$ of chart transformations for a chart $h_\alpha \in \mathfrak{A}$, and differentiability is a local property. As an element in the family of differentiable atlases, the atlas $\mathfrak{D}$ can obviously not be enlarged by the addition of further charts, and it is the largest differentiable atlas which contains $\mathfrak{A}$. Thus each differentiable atlas unequivocally determines a maximal differentiable atlas $\mathfrak{D}(\mathfrak{A})$, so that $\mathfrak{A} \subset \mathfrak{D}(\mathfrak{A})$; and $\mathfrak{D}(\mathfrak{A}) = \mathfrak{D}(\mathfrak{B})$ if and only if the atlas $\mathfrak{A} \cup \mathfrak{B}$ is differentiable. We formulate:

(1.4) Definition. A *differentiable structure* on a topological manifold is a maximal differentiable atlas. A *differentiable manifold* is a topological manifold, together with a differentiable structure.

In order to specify a differentiable structure on a manifold, one must specify a differentiable atlas and, in general, one will clearly not choose the maximal one, but preferably one as small as possible.

Henceforth we shall implicitly assume that all charts and atlases of a differentiable manifold with a differentiable structure $\mathfrak{D}$ are contained in $\mathfrak{D}$. In the notation, as usual, we employ the abbreviated form M, and not $(M, \mathfrak{D})$ for a differentiable manifold.

(1.5) Examples. (a) If $U \subset \mathbb{R}^n$ is an open subset, then the atlas $\{\mathrm{Id}_U\}$, which only contains the single chart $\mathrm{Id}: U \to U$, defines the usual differentiable structure. Furthermore, every homeomorphism $h: U \to U$ defines a differentiable atlas $\{h\}$, which gives the same differentiable structure if and only if h is a diffeomorphism. On an open subset of $\mathbb{R}^n$, one can therefore easily describe various differentiable structures for $n > 0$. However, as we shall yet see, using such atlases with only one chart $h: U \to U$, one does not obtain substantially different differentiable manifolds.

(b) The sphere $S^n = \{x \in \mathbb{R}^{n+1} \mid |x| := \sqrt{(x_1^2 + \ldots + x_{n+1}^2)} = 1\}$ possesses a differentiable atlas whose differentiable structure we shall always consider as introducing the standard structure on S^n. The chart domains are the sets

$$U_{kj} = \{x \in S^n \mid (-1)^j x_k > 0\},$$

the charts are

$$h_{kj}: U_{kj} \to \mathring{D}^n \text{ (the open solid ball)}$$

$$x \mapsto (x_1, \ldots, x_{k-1}, x_{k+1}, \ldots, x_{n+1}), \text{ see Fig. 5,}$$

so that the chart h_{kj} forgets the kth coordinate. It is easy to verify that this atlas is differentiable, since the map $h_{kj}^{-1}: \mathring{D}^n \to S^n$ has the kth coordinate (missing in $\mathring{D}^n$) $(-1)^j (1 - \Sigma_{i \neq k} x_i^2)^{1/2}$, which is clearly a differentiable function on $\mathring{D}^n$ in the usual sense; and h_{kj} results by restricting a differentiable mapping $\mathbb{R}^{n+1} \to \mathbb{R}^n$.

(c) The real projective space $\mathbb{R}\mathbf{P}^n$ is the quotient space of the sphere S^n under the equivalence relation defined by $x \sim -x$. A point $p \in \mathbb{R}\mathbf{P}^n$ is described by

$$p = [x] = [x_0, \ldots, x_n] = [-x_0, \ldots, -x_n], \quad \sum_{i=0}^{n} x_i^2 = 1,$$

and the equivalence relation identifies precisely the subsets $U_{k,0}$ and $U_{k,1}$ of the sphere. Therefore, the subsets

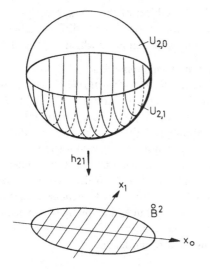

Fig. 5

$$U_k = \{[x] \in \mathbb{R}\mathbf{P}^n \mid x_k \neq 0\}$$

are open in $\mathbb{R}\mathbf{P}^n$, and one has charts

$$h_k \colon U_k \to \mathring{D}^n, \quad [x_0, \ldots, x_n] \mapsto x_k \cdot |x_k|^{-1} \cdot (x_0, \ldots, x_{k-1}, x_{k+1}, \ldots, x_n)$$

for a differentiable atlas.

The projective spaces are examples of differentiable manifolds which arise naturally as abstract manifolds and not as subsets of Euclidean space. And, initially, it is not obvious that a projective space is homeomorphic to a subset of Euclidean space. One also obtains the topological manifold $\mathbb{R}\mathbf{P}^n$ when one identifies antipodal boundary points of the ball $D^n = \{x \in \mathbb{R}^n \mid |x| \leq 1\}$, that is, forms the quotient for the equivalence relation '$x \sim -x$ for $|x| = 1$'. In this way, one can visualise the projective plane $\mathbb{R}\mathbf{P}^2$ as the result from glueing together a Möbius band B and a disc $A \cup C$ along their common boundary S^1, as in Fig. 6.

(d) An open subset of a differentiable manifold possesses a natural structure as a differentiable manifold.

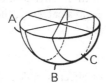

Fig. 6

Differentiable manifolds will be the subject of this book, more precisely, the category of differentiable manifolds. Its 'objects' are differentiable manifolds; its 'morphisms' are the differentiable mappings which we now define.

(1.6) Definition. A continuous mapping $f: M \to N$ between differentiable manifolds is termed *differentiable at the point $p \in M$* if for some (and therefore for every) chart $h: U \to U'$, $p \in U$ and $k: V \to V'$, $f(p) \in V$ of M and N respectively, the composition $k \cdot f \cdot h^{-1}$ is differentiable at the point $h(p) \in U'$. Note that this mapping is defined in the neighbourhood $h(f^{-1}V \cap U)$ of $h(p)$, see Fig. 7. The mapping f is termed differentiable if it is differentiable at every point $p \in M$. In other words: one knows when one can call a mapping between chart domains of M and N differentiable, because these are identified by the charts with open subsets of Euclidean space, and locally a continuous mapping is always written as a mapping between chart domains. Independence from the particular choice of chart depends upon the fact that the chart transformations are differentiable.

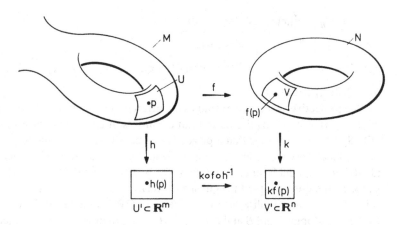

Fig. 7

Remark and notation. The identity mapping of a differentiable manifold is differentiable; the composition of differentiable mappings is differentiable. One assumes both these assertions in saying that differentiable manifolds and mappings form a category, the differentiable category which will be written C^∞ for short.

Correspondingly, let

$C^\infty(M, N) :=$ the set of differentiable maps $M \to N$;

$C^\infty(M) := C^\infty(M, \mathbb{R})$.

The composition of differentiable maps is therefore a map

$$C^\infty(M, N) \times C^\infty(L, M) \vec{\to} C^\infty(L, N), \quad (f, g) \mapsto f \cdot g.$$

Many concepts arise in a category in a purely formal way; they are formulated using the maps of the category and their composition, as, for example, *isomorphism*, *sum*, and *product*.

(1.7) Definition. A *diffeomorphism* is an invertible differentiable map.

'Invertible', it is worth noting, means invertible in the differentiable category, therefore $f: M \to N$ is a diffeomorphism if there is a differentiable map $g: N \to M$, so that $f \cdot g = \mathrm{Id}_N$ and $g \cdot f = \mathrm{Id}_M$. This means, in other words: f is bijective and, also, f^{-1} is differentiable. We denote diffeomorphisms by '$\cong$'; they form the isomorphisms of the differentiable category.

A differentiable homeomorphism need not be a diffeomorphism, as is shown by the map $\mathbb{R} \to \mathbb{R}, x \mapsto x^3$.

For example, in (1.5(a)) we have introduced, in general, many distinct differentiable structures on an open subset $U \subset \mathbb{R}^n$, but the differentiable manifolds U with atlas $\{\mathrm{Id}\}$, and U with atlas $\{h\}$, are of course diffeomorphic; $h: U \to U$ is a diffeomorphism $(U, \{h\}) \to (U, \{\mathrm{Id}\})$ of the second onto the first. Thus, both manifolds are essentially the same in so far as their differential topology is concerned.

In contrast, the problem of constructing two distinct differentiable structures on a topological manifold, so that the resulting differentiable manifolds are not diffeomorphic, is very deep indeed. For example, the topological 7-sphere possesses exactly 15 mutually distinct non-diffeomorphic structures as a differentiable manifold. These are precisely 15 mutually distinct differentiable manifolds which are, however, all homeomorphic to the sphere S^7 (Kervaire & Milnor, 1963). Such results are far beyond the scope of this book.

Every chart $h: U \to U'$ of M is a diffeomorphism between U and U', where U' carries the standard structure as an open subset of $\mathbb{R}^n$ (1.5(d)), and the differentiable structure of M consists precisely of the set of all diffeomorphisms of open subsets of M with open subsets of $\mathbb{R}^n$.

The function $t \mapsto \tan((\pi/2)t)$ defines a diffeomorphism $(-1, 1) \to \mathbb{R}$.

Differential topology deals with those properties which remain constant under the action of diffeomorphisms. For local considerations, one can therefore always assume that one is dealing with an open subset of $\mathbb{R}^n$; instead of a function f on U, one considers $f \cdot h^{-1}$ on U'; instead of an open subset $V \subset U$, the subset $h(V) \subset U'$; and so forth. Since images in $\mathbb{R}^n$ are given by their coordinates, one also often describes a chart of M around p in terms of a local coordinate system. The chart $h: U \to U'$ is written in components as $h = (h_1, \ldots, h_n)$, where the coordinate functions $h_i: U \to \mathbb{R}$ are differentiable functions; by translation in $\mathbb{R}^n$, one can further assume that $h(p) = 0 = (0, \ldots, 0)$ for a fixed point $p \in U$. Thus, in a neighbourhood U of p, after the introduction of a coordinate system, every point is uniquely determined by the values of the coordinate functions. Thus, for each point in U, one can assign coordinates

with
$$(x_1, \ldots, x_n)$$
$$(0, \ldots, 0) = \text{coordinates of } p.$$

A function on U is thus differentiable if and only if it is differentiable as a function of the coordinates in the usual meaning of the differential calculus.

In the differentiable category there are sums and products:

(1.8) Definition. The disjoint union of two n-dimensional differentiable manifolds M_1, M_2 is, in a natural way, a differentiable manifold expressed by $M_1 + M_2$, see Fig. 8. The topology is determined by the fact that both manifolds M_1, M_2 are open subsets of $M_1 + M_2$, and a differentiable atlas is the union of atlases of both manifolds.

The manifold $M_1 + M_2$ is called the *(differentiable) sum* of M_1 and M_2. One has canonical inclusions

$$i_\nu : M_\nu \to M_1 + M_2$$

as open subsets. A map $f : M_1 + M_2 \to N$ is then clearly differentiable if and only if both restrictions $f \circ i_\nu$ are differentiable; in other words one has a canonical bijection

$$C^\infty(M_1 + M_2, N) \to C^\infty(M_1, N) \times C^\infty(M_2, N), \quad f \mapsto (f \circ i_1, f \circ i_2)$$

for every differentiable manifold N (*universal property of the sum*).

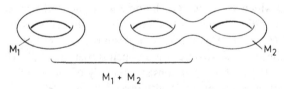

Fig. 8

Dually, one constructs the Cartesian product $M_1 \times M_2$ of two differentiable manifolds M_1, M_2 of dimensions n, k, and gives this the structure of a $(n + k)$-dimensional differentiable manifold which is called the *(differentiable)product* of M_1 and M_2. If $h_\nu : U_\nu \to U'_\nu$ are charts of the differentiable structure of M_ν, then

$$h_1 \times h_2 : U_1 \times U_2 \to U'_1 \times U'_2 \subset \mathbb{R}^n \times \mathbb{R}^k = \mathbb{R}^{n+k}$$

is a chart of $M_1 \times M_2$, and the set of all these charts defines the differentiable structure for $M_1 \times M_2$ (see Fig. 9). ($M_1 = M_2 = S^1$, with (p, q) a general point in the product.) One has canonical projections $p_\nu : M_1 \times M_2 \to M_\nu$

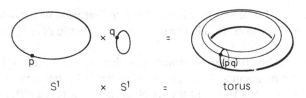

S¹ × S¹ = torus

Fig. 9

and, analogously to the sum, a canonical bijection

$$C^\infty(N, M_1 \times M_2) \to C^\infty(N, M_1) \times C^\infty(N, M_2), \ f \mapsto (p_1 \circ f, p_2 \circ f)$$

for every differentiable manifold N (*universal property of the product*). The last remark states that a map into the product is differentiable if and only if both its components $f_\nu = p_\nu \circ f$ are differentiable; locally one maps into a chart domain $U_1 \times U_2$, and the composition with a chart

$$h_1 \times h_2 : U_1 \times U_2 \to U_1' \times U_2' \subset \mathbb{R}^{n+k}$$

is then differentiable if and only if both its components are differentiable.

Less canonical and, therefore, not so uniformly defined in the literature, is the concept of submanifold.

(1.9) Definition. A subset $N \subset M^{n+k}$ is called an *n-dimensional differentiable submanifold* of M if, for every point $p \in N$, there exists a chart around p

$$h: U \to U' \subset \mathbb{R}^{n+k} = \mathbb{R}^n \times \mathbb{R}^k$$

so that

$$h(N \cap U) = U' \cap \mathbb{R}^n$$

where we consider $\mathbb{R}^n$ as $\mathbb{R}^n \times 0 \subset \mathbb{R}^n \times \mathbb{R}^k$.

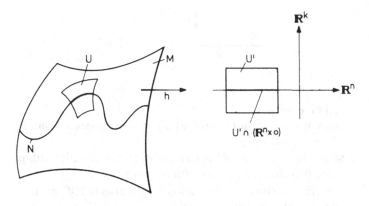

Fig. 10

The number $k = \dim M - \dim N$ is called the *codimension* of the submanifold. In short one says: locally the submanifold N lies in M as $\mathbb{R}^n$ lies in $\mathbb{R}^{n+k}$.

The definition is justified by the remark that there is a canonical differentiable structure on N. From a chart h, as in definition (1.9), one obtains a chart $h' = h \,|\, U \cap N \to U' \cap \mathbb{R}^n$, and the set of all these charts is a differentiable atlas for N, see Fig. 10.

(1.10) Definition. A differentiable map $f: N \to M$ is called an *embedding* if $f(N) \subset M$ is a differentiable submanifold, and $f: N \to f(N)$ is a diffeomorphism.

If N and M have the same dimension, then $f(N)$ is open in M, as definition (1.9) unmistakably shows, and the inclusion of an open subset is also an embedding. Otherwise, it is necessary that $\dim N < \dim M$. Every point $p \in M$ defines an embedding

$$i_p: N \to M \times N, \quad q \mapsto (p, q)$$

so that $p_2 \circ i_p = \mathrm{Id}_N$ and, similarly, every point $p \in M$ defines a projection $\pi_p: M + N \to M$, so that $\pi_p \circ i_1 = \mathrm{Id}_M$. The second factor, of course, behaves quite analogously; if $p \in M$ and $q \in N$, then $i_p(N)$ and $i_q(M)$ meet precisely in the point $(p, q) \in M \times N$, see Fig. 11.

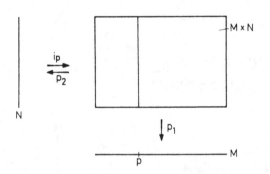

Fig. 11

(1.11) Exercises

1 Show that every (differentiable) manifold possesses a countable (differentiable) atlas.
2 Show that the sphere S^n possesses a differentiable atlas with precisely two charts. Also, one with only one chart?
3 Describe the chart transformation for the atlas of $\mathbb{R}P^n$ in (1.5(c)), and show that it is differentiable.

4 Let M be a differentiable manifold and $\tau: M \to M$ a fixed point free involution, that is, τ is a diffeomorphism with $\tau \circ \tau = \mathrm{Id}_M$ and $\tau(x) \neq x$ for all x.
  Show that the quotient space M/τ, which is obtained from M by identification of points corresponding to each other under τ, is a topological manifold which possesses exactly one differentiable structure with respect to which the projection $M \to M/\tau$ is locally diffeomorphic.

5 Show that $\mathbb{R}P^1 \cong S^1$.

6 Provide the surface of a cube $\{x \in \mathbb{R}^{n+1} | \max \{|x_i|\} = 1\}$ with the structure of a differentiable manifold.

7 Let M be a differentiable manifold and $f: N \to M$ a homeomorphism. Prove that N possesses exactly one structure as a differentiable manifold, so that f is diffeomorphic.

8 Provide the complex projective space $\mathbb{C}P^n$ with the structure of a $2n$-dimensional differentiable manifold. This space is defined as follows: on the complex vector space $\mathbb{C}^{n+1}$, one has the equivalence relation $x \sim y \Leftrightarrow$ there is a number $\lambda \in \mathbb{C}, \lambda \neq 0$, so that $\lambda x = y$. The quotient space $(\mathbb{C}^{n+1} - \{0\})/\sim$ is defined to be $\mathbb{C}P^n$.

9 Prove that if M is a non-empty, n-dimensional manifold and $k \leqslant n$, then there is an embedding $\mathbb{R}^k \to M$.

10 Let N be a compact, M a connected manifold, both of dimension n and non-empty. Let $f: N \to M$ be an embedding. Show that f is a diffeomorphism.

11 Show that S^n is a submanifold of $\mathbb{R}^{n+1}$.

12 Describe an embedding $S^1 \times S^1 \to \mathbb{R}^3$ by means of elementary functions.

13 Show that the composition of two embeddings is again an embedding and that the Cartesian product $f_1 \times f_2: N_1 \times N_2 \to M_1 \times M_2$, of two embeddings f_1, f_2, is again an embedding.

14 Show that if the n-dimensional manifold M is a product of spheres, then there exists an embedding $M \to \mathbb{R}^{n+1}$.
  Hint: describe an embedding $S^n \times \mathbb{R} \to \mathbb{R}^{n+1}$ and use 13.

15 The points of $\mathbb{C}P^k$ (see 8) are described by the homogeneous co-ordinates $x = [x_0, \ldots, x_k] := $ class of $(x_0, \ldots, x_k)$ under $\sim$. Show that the mapping

$$f: \mathbb{C}P^m \times \mathbb{C}P^n \to \mathbb{C}P^{mn+m+n}$$

$$(x, y) \mapsto [x_0 y_0, x_0 y_1, \ldots, x_\nu y_\mu, \ldots, x_m y_n]$$

is an embedding. Show the same for the real projective spaces.

16 Let $M(m \times n)$ be the vector space of real $(m \times n)$-matrices, and $M_r(m \times n)$ the subset of matrices of rank r. Then $M_r(m \times n)$ is a

submanifold of $M(m \times n)$ of codimension $(n-r) \cdot (m-r)$, for $r \leqslant \min\{m, n\}$.

Hint: a typical chart domain around a point of $M_r(m \times n)$ is given by the set $U \subset M(m \times n)$ of matrices of the form

$$\begin{pmatrix} A & AB \\ D & DB + C \end{pmatrix}, \quad A \in M(r \times r), \; \det(A) \neq 0.$$

Such a matrix lies in $M_r(m \times n)$ if and only if $C = 0$.

17 The inclusion $\mathbb{R}^{n+1} \subset \mathbb{R}^{n+2}$ induces an embedding $\mathbb{RP}^n \subset \mathbb{RP}^{n+1}$ and $\mathbb{RP}^{n+1} - \mathbb{RP}^n \cong \mathbb{R}^{n+1}$.

18 Let $\mathbb{R}^{n+1} = \{(x, a_0, \ldots, a_{n-1}) | x, a_i \in \mathbb{R}\}$. The set of points such that $x^n + a_{n-1}x^{n-1} + \ldots + a_0 = 0$ is a submanifold of codimension 1 of $\mathbb{R}^{n+1}$, and is diffeomorphic to $\mathbb{R}^n$.

19 The set $C^\infty(M)$ is an algebra under the natural addition and multiplication of functions. A differentiable mapping $f \colon M \to N$ defines an algebra homomorphism

$$f^* \colon C^\infty(N) \to C^\infty(M), \quad \phi \mapsto \phi \circ f$$

with the functorial properties: $\mathrm{Id}_M^* = \mathrm{Id}; (f \circ g)^* = g^* \circ f^*$.

20 Notation as in 19. For a point $p \in M$ let

$$\mathfrak{M}_p = \{\phi \in C^\infty(M) | \phi(p) = 0\}.$$

Show:

(a) $\mathfrak{M}_p$ is a maximal ideal of $C^\infty(M)$.

(b) If M is compact and $\mathfrak{M} \in C^\infty(M)$ is a maximal ideal, then there exists some $p \in M$ such that $\mathfrak{M} = \mathfrak{M}_p$.

2
Tangent space

Problems in differential topology often divide into a local and a global part; in this section we begin explaining basic local concepts.

The dominating concept of local theory is that of the tangent space at a point $p \in M$ of a manifold. Let us assume that the manifold is embedded in Euclidean space $\mathbb{R}^n$, then it is quite obvious that to every point $p \in M$ there is assigned a certain linear subspace of $\mathbb{R}^n$, the space of tangent vectors of M at p, the velocity vectors of possible movements on M. Thus the sphere S^n is embedded in $\mathbb{R}^{n+1}$ as $S^n = \{x \in \mathbb{R}^{n+1} | \, |x| = 1\}$, and the tangent space at the point $x \in S^n$ is the set of vectors $\{v \in \mathbb{R}^{n+1} | \langle v, x \rangle = 0\}$, see Fig. 12. Since, in general, such an embedding is not canonically given, we must describe the tangent space by the intrinsic properties of the manifold.

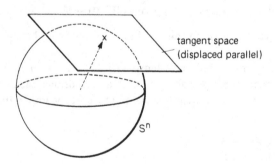

Fig. 12

For local considerations it is clear that one takes into account not just maps $f: M \to N$ defined on all of M, but also those maps which are defined only in a neighbourhood of $p \in M$. Two such maps can be looked upon as equal if they agree in a (perhaps smaller) neighbourhood. Thus on the set of differentiable maps

$$\{f \, | \, f: U \to N, \text{ for a neighbourhood } U \text{ of } p \in M\}$$

we construct the following equivalence relation:

$f \sim g \Leftrightarrow$ there is a neighbourhood V of p, so that $f \mid V = g \mid V$.

(2.1) Definition. An equivalence class for this relation is called the *germ* of a map $M \to N$ at p. We denote such a germ, represented by f, as $\bar{f}: (M,p) \to N$ and also $\bar{f}: (M,p) \to (N,q)$, if $f(p) = q$. Given germs $(M,p) \underset{\bar{f}}{\to} (N,q) \underset{\bar{g}}{\to} (L,r)$, one obtains a *composition* $\bar{g} \cdot \bar{f}: (M,p) \to (L,r)$ as the germ of the composition of suitable representatives. A *function germ* is a differentiable germ $(M,p) \to \mathbb{R}$. The set of all function germs around $p \in M$ is written as $\mathscr{E}(p)$.

Thus $\mathscr{E}(p)$ has the structure of a real algebra: addition and multiplication are defined by the corresponding operations on representatives. A differentiable germ $\bar{f}: (M,p) \to (N,q)$ defines by composition a homomorphism of algebras

$$f^*: \mathscr{E}(q) \to \mathscr{E}(p), \quad \bar{\phi} \mapsto \bar{\phi} \cdot \bar{f},$$

and one has the functorial properties

$$\mathrm{Id}^* = \mathrm{Id}; \quad (g \cdot f)^* = f^* \cdot g^*.$$

From the functorial properties it follows in particular that an invertible germ $\bar{f}$ relative to composition induces an isomorphism f^*:

$$\bar{f} \cdot \bar{f}^{-1} = \mathrm{Id} \Rightarrow f^{-1*} \cdot f^* = \mathrm{Id}.$$

If, therefore, $p \in M^n$, then one can find a chart h about p, which defines an invertible germ $\bar{h}: (M,p) \to (\mathbb{R}^n, 0)$, and therefore an isomorphism

$$h^*: \mathscr{E}_n \to \mathscr{E}(p); \quad \mathscr{E}_n = \text{set of germs } (\mathbb{R}^n, 0) \to \mathbb{R}.$$

The study of the algebras $\mathscr{E}(p)$ can thus be limited to the typical examples $\mathscr{E}_n$.

Since we have thus far directed our attention to the local, we now turn to tangent spaces. There are three prevailing equivalent definitions, each of which has its advantages, and we wish to learn to move freely among them: the definitions

    (A) the algebraist's
    (Ph) the physicist's
    (G) the geometer's

(2.2) The algebraist's definition. The *tangent space* T_pM of the differentiable manifold M at point p is the real vector space of the derivations of $\mathscr{E}(p)$. A derivation of $\mathscr{E}(p)$ is a linear map $X: \mathscr{E}(p) \to \mathbb{R}$ which satisfies the product rule

$$X(\bar{\phi} \cdot \bar{\psi}) = X(\bar{\phi}) \cdot \bar{\psi}(p) + \bar{\phi}(p) \cdot X(\bar{\psi}).$$

A differentiable germ $\bar{f}: (M,p) \to (N,q)$, for example one associated with a differentiable map $f: M \to N$, induces the algebra homomorphism

$f^*: \mathcal{E}(q) \to \mathcal{E}(p)$ and thereby the linear *tangent map* (the *differential*) of f at p:

$$T_p f: T_p M \to T_q N$$

$$X \mapsto X \cdot f^*.$$

One immediately checks, that a linear combination of derivations is again a derivation, that these thus form a vector space. From the product rule it follows that $X(1) = X(1) + X(1)$, therefore $X(1) = 0$ for the constant function with value 1, thus, because of linearity, $X(c) = 0$ also for every constant c. The definition of the differential implies that for a germ $\bar{\phi}: (N, q) \to \mathbb{R}$:

$$T_p f(X)(\bar{\phi}) = X \cdot f^*(\bar{\phi}) = X(\phi \cdot f).$$

From this, or from the functorial properties of *, if follows that for a composition $(M, p) \underset{f}{\to} (N, q) \underset{g}{\to} (L, r)$, one has the functorial property $T_p(\bar{g} \cdot \bar{f}) = T_q \bar{g} \cdot T_p \bar{f}$ for the tangential map. One reads directly from the definition that the tangential map is linear.

Now if $\bar{h}: (N, p) \to (\mathbb{R}^n, 0)$ is the germ of a chart, then the induced map $h^*: \mathcal{E}_n \to \mathcal{E}(p)$ is an isomorphism, as is the tangential map $T_p h: T_p N \to T_0 \mathbb{R}^n$. In order to describe the latter vector space the following is useful:

(2.3) Lemma. *Let U be an open ball around the origin of $\mathbb{R}^n$ or $\mathbb{R}^n$ itself, and $f: U \to \mathbb{R}$ a differentiable function, then there exist differentiable functions $f_1, \dots, f_n: U \to \mathbb{R}$, so that*

$$f(x) = f(0) + \sum_{\nu=1}^{n} x_\nu \cdot f_\nu(x).$$

Proof.

$$f(x) - f(0) = \int_0^1 \frac{\mathrm{d}}{\mathrm{d}t} f(tx_1, \dots, tx_n)\,\mathrm{d}t = \sum_{\nu=1}^{n} x_\nu \int_0^1 D_\nu f(tx_1, \dots, tx_n)\,\mathrm{d}t,$$

where D_ν denotes the partial derivative with respect to the νth variable. Therefore, set

$$f_\nu(x) := \int_0^1 D_\nu f(tx_1, \dots, tx_n)\,\mathrm{d}t. \qquad \square$$

Among the derivations – as the name implies – of the algebra $\mathcal{E}_n$ are the partial derivatives, which we usually write in the old fashioned way:

$$\frac{\partial}{\partial x_\nu}: \mathcal{E}_n \to \mathbb{R}, \quad \bar{\phi} \mapsto \frac{\partial}{\partial x_\nu} \phi(0).$$

Consequence. The $\partial/\partial x_\nu$, $\nu = 1, \ldots, n$, form a basis of the vector space $T_0 \mathbb{R}^n$ of the derivations of $\mathscr{E}_n$.

Proof. If the derivation $\sum_{\nu=1}^{n} a_\nu(\partial/\partial x_\nu) = 0$ then, in particular, one obtains for $\bar{x}_\mu$, the μth coordinate function: $a_\mu = \sum_{\nu=1}^{n} a_\nu(\partial \bar{x}_\mu/\partial x_\nu) = 0$ for all μ. Therefore the $\partial/\partial x_\nu$ are linearly independent.

Now let $X \in T_0(\mathbb{R}^n)$, $X(\bar{x}_\nu) =: a_\nu$, then we shall show that:

$$X = \sum_{\nu=1}^{n} a_\nu \frac{\partial}{\partial x_\nu}.$$

If we set $Y := X - \sum_{\nu=1}^{n} a_\nu(\partial/\partial x_\nu)$, then Y is a derivation and, by construction, $Y(\bar{x}_\nu) = 0$ for every coordinate function. If $\bar{f} \in \mathscr{E}_n$ is an arbitrary function germ, we then write, by lemma (2.3), $\bar{f} = \bar{f}(0) + \sum_{\nu=1}^{n} \bar{x}_\nu \cdot \bar{f}_\nu$ and obtain

$$Y(\bar{f}) = Y(f(0)) + \sum_{\nu=1}^{n} Y(\bar{x}_\nu) \cdot f_\nu(0) = 0. \qquad \square$$

At this point, note that the tangent space at a point of an n-dimensional differentiable manifold has the vector space dimension n, so that the dimension is indeed unequivocally defined. It is not so easy to see this in the topological case, but it is nonetheless true.

After introducing local coordinates $(x_1, \ldots, x_n)$ about a point $p \in N^n$, we can explicitly describe the vectors in $T_p N$ as linear combinations of the $\partial/\partial x_i$. If $\bar{f}: (N^n, p) \to (M^m, q)$ is a differentiable germ, and if we also construct local coordinates $(y_1, \ldots, y_m)$ around q, then $\bar{f}$ is written as a germ $(\mathbb{R}^n, 0) \to (\mathbb{R}^m, 0)$, which we shall also simply denote by $\bar{f}$:

$$
\begin{array}{ccc}
(N, p) & \xrightarrow{\bar{f}} & (M, q) \\
\downarrow & & \downarrow \\
(\mathbb{R}^n, 0) & \xrightarrow{\bar{f}} & (\mathbb{R}^m, 0)
\end{array}
$$

The tangential map $T_0 \bar{f}$ is computed as follows: If $\bar{\phi} \in \mathscr{E}_m$, then following definition (2.2) and the chain rule,

$$T_0 \bar{f} \left(\frac{\partial}{\partial x_i} \right) (\bar{\phi}) = \frac{\partial}{\partial x_i} (\bar{\phi} \cdot \bar{f}) = \sum_{j=1}^{m} \frac{\partial \phi}{\partial y_j}(0) \cdot \frac{\partial f_j}{\partial x_i}(0),$$

therefore

$$T_0 \bar{f} \left(\frac{\partial}{\partial x_i} \right) = \sum_{j=1}^{m} \frac{\partial f_j}{\partial x_i}(0) \frac{\partial}{\partial y_j}.$$

The matrix

$$Df := \left(\frac{\partial f_i}{\partial x_j} \right)$$

is called the *Jacobi matrix*. We can therefore compute the differential of $\bar{f}$ in matrix notation thus: if $v = \Sigma\, a_i(\partial/\partial x_i)$, then $T_0\bar{f}(v) = \Sigma\, b_j(\partial/\partial y_j)$, where

$$b = Df_0 \cdot a.$$

We summarise all this as:

(2.4) Theorem. *If one introduces local coordinates $(x_1, \ldots, x_n)$ and $(y_1, \ldots, y_m)$ around $p \in N^n$ and $q \in M^m$ respectively, then the derivations $\partial/\partial x_i$, $\partial/\partial y_j$ form vector space bases of T_pN and T_qM respectively, and the tangential map of a germ $\bar{f}: (N, p) \to (M, q)$ with respect to these bases is given by*

$$Df_0 \colon \mathbb{R}^n \to \mathbb{R}^m. \qquad \qquad \square$$

The definition of the algebraist is the easiest to apply. However, it is rather abstract (and also unsuitable, when one considers infinite dimensional manifolds, or just finitely often differentiable ones).

Physicists proceed from the coordinate dependent version of theorem (2.4). One hears descriptions such as: 'A contravariant vector or tensor of the first order is a real n-tuple which transforms according to the Jacobi matrix'. This we interpret as follows: if $\bar{h}, \bar{k} \colon (N, p) \to (\mathbb{R}^n, 0)$ are germs of charts, then the chart transformation $\bar{g} := \bar{k} \cdot \bar{h}^{-1} \colon (\mathbb{R}^n, 0) \to (\mathbb{R}^n, 0)$ is an invertible differentiable germ. The various invertible germs $(\mathbb{R}^n, 0) \to (\mathbb{R}^n, 0)$, that is, all possible chart transformations, form a group $\mathscr{G}$ under composition '$\cdot$', and thus for two chart germs $\bar{h}, \bar{k}$ there is exactly one $\bar{g} \in \mathscr{G}$, so that $\bar{g} \cdot \bar{h} = \bar{k}$. To every $\bar{g} \in \mathscr{G}$ we assign the Jacobi matrix at the origin Dg_0 and, as in the differential calculus, the product of the matrices is then associated with the composition of the maps; in particular, one has a homomorphism of groups

$$\mathscr{G} \to GL\,(n, \mathbb{R}), \quad \bar{g} \mapsto Dg_0$$

from $\mathscr{G}$ into the linear group of invertible matrices.

(2.5) The physicist's definition. A *tangent vector* at the point $p \in N^n$ is a rule assigning to each chart germ $\bar{h} \colon (N, p) \to (\mathbb{R}^n, 0)$ a vector $v = (v_1, \ldots, v_n) \in \mathbb{R}^n$, so that the vector $Dg_0 \cdot v$ corresponds to the chart germ $\bar{g} \cdot \bar{h}$, see Fig. 13.

Thus if we denote by K_p the set of chart germs

$$\bar{h} \colon (N, p) \to (\mathbb{R}^n, 0),$$

the physicist's tangent space $T_p(N)_{Ph}$ equals the set of maps

$$v \colon K_p \to \mathbb{R}^n,$$

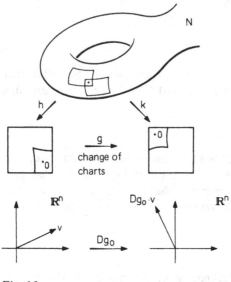

Fig. 13

for which

$$v(\bar{g} \cdot \bar{h}) = Dg_0 \cdot v(\bar{h}) \text{ for all } \bar{g} \in \mathscr{G}.$$

These maps form a vector space because Dg_0 is a linear map. For a fixed chart h, clearly one can arbitrarily choose the vector $v \in \mathbb{R}^n$, and the choice of all other chart germs is fixed by this: the vector space $T_p(N)_{Ph}$ is isomorphic to $\mathbb{R}^n$. An isomorphism is given by the choice of a local coordinate system. The canonical isomorphism

$$T_p N \to T_p(N)_{Ph}$$

with the algebraically defined tangent space (2.2), given the chart $\bar{h} = (\bar{h}_1, \dots, \bar{h}_n): (N, p) \to (\mathbb{R}^n, 0)$ assigns to the derivation $X \in T_p N$ the vector $(X(\bar{h}_1), \dots, X(\bar{h}_n)) \in \mathbb{R}^n$. The components of this vector are precisely the coefficients of X with respect to the basis of $T_p N$ in (2.4); through identification they are transformed by the Jacobi matrix because the basis in (2.4) is mapped by the transposed Jacobi matrix.

The differential is, given local coordinate systems about the image and pre-image points, described by the Jacobi matrix, as in (2.4), although formally this is rather awkward to write down because of the many coordinate systems.

The definition of the geometer is the most intuitive one; it is derived from the concept that the tangent vectors are velocity vectors of paths through the point p at this point. Everything is of course again considered locally near the point.

(2.6) **The geometer's definition.** On the set W_p of germs of differentiable maps

$$\bar{w}: (\mathbb{R}, 0) \to (N, p)$$

(that is, the germs of paths passing through p) we formulate the equivalence relation $\bar{w} \sim \bar{v}: \Leftrightarrow$ for every function germ $\bar{f} \in \mathscr{E}(p)$, $(d/dt)\bar{f} \circ \bar{w}(0) = (d/dt)\bar{f} \circ \bar{v}(0)$. An equivalence class $[w]$, for this relation, is a tangent vector to the point p, see Fig. 14.

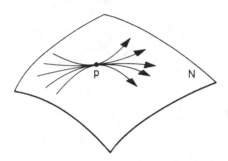

Fig. 14

Two path germs define the same tangent vector if and only if they define the same 'differentiation of functions in the direction of the curve'. To every equivalence class $[w]$ there is, in this way uniquely associated, the derivation X_w of $\mathscr{E}(p)$:

$$X_w(\bar{f}) := \frac{\mathrm{d}}{\mathrm{d}t} \bar{f} \circ \bar{w}(0).$$

This association defines an injective map

$$W_p/\sim := T_p(N)_G \to T_pN, \quad [w] \mapsto X_w$$

of the set of equivalence classes of path germs into the tangent space. This map is also surjective since, if (in local coordinates) $w(t) = (ta_1, \ldots, ta_n)$, then $X_w = \Sigma_{\nu=1}^n a_\nu(\partial/\partial x_\nu)$. Indeed, one only needs to check an equality of derivations $X_w = X_v$ on the coordinate functions of a local coordinate system (the values are precisely the coefficients with respect to the basis $\partial/\partial x_\nu$). Hence one can also say: $w \sim v$ if and only if for a local coordinate system $(d/dt)w_i(0) = (d/dt)v_i(0)$ for $i = 1, \ldots, n$.

In this definition the tangent map is also very clear: a germ $\bar{f}: (N, p) \to (M, q)$ induces the map

$$T_p(N)_G \to T_q(M)_G, \quad [w] \mapsto [f \circ w], \text{ see Fig. 15.}$$

The fact that this definition is compatible with the earlier definition (2.2) is shown by the equation

$$X_{fw}(\bar{\phi}) = \frac{\mathrm{d}}{\mathrm{d}t} \bar{\phi}\bar{f}\bar{w}(0) = X_w(\bar{\phi}f) = T_pf(X_w)(\bar{\phi}).$$

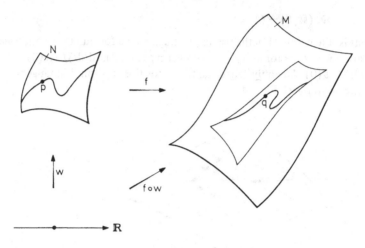

Fig. 15

From now on we shall make no distinction between the different definitions of the tangent space. Our intuition follows the geometrical definition; explicit computations, where necessary, use the coordinate description (2.4).

A finite dimensional real vector space V is a differentiable manifold. A choice of basis determines an isomorphism $V \xrightarrow{\approx} \mathbb{R}^n$, which one can take as a chart for an atlas. Because linear maps of $\mathbb{R}^n$ are differentiable, the differentiable structure defined in this way is independent of the basis. The tangent space $T_p V$ is canonically isomorphic to V for each point $p \in V$. One can describe the isomorphism thus: the curve $w_v \colon t \mapsto p + tv$ through p corresponds to the vector $v \in V$, and $[w_v]$ is the associated tangent vector (geometer's language). Naturally, if M has dimension n, the tangent space $T_p M$ is always isomorphic to $\mathbb{R}^n$ but, in general, there is no canonical, in some way preferred, isomorphism. This we shall see even more clearly in the next section.

(2.7) Exercises

1 Show that $\mathfrak{m}(p) := \{\bar{\phi} \in \mathscr{E}(p) \mid \bar{\phi}(p) = 0\}$ is the only maximal ideal of $\mathscr{E}(p)$.

2 Show that if $p \in M^n$ and $n \neq 0$, then the ideal $\mathfrak{m}(p)$ in exercise 1 is not the only ideal $\neq 0$, $\mathscr{E}(p)$ of $\mathscr{E}(p)$.

3 Show that if $f \colon M \to N$ is an embedding and $f(p) = q$, then the map $f^* \colon \mathscr{E}(q) \to \mathscr{E}(p)$ is surjective and $T_p(f)$ injective.

4 Show that the maximal ideal $\mathfrak{m}_n \subset \mathscr{E}_n$ is generated by the germs $\bar{x}_1, \ldots, \bar{x}_n$ of the coordinate functions.

5 Show that if $\mathfrak{m}_n \subset \mathscr{E}_n$ is the maximal ideal, then $\mathfrak{m}_n^k$ is the ideal of the germs $\bar{f}$, for which all partial derivatives of order $< k$ vanish at the origin.

6 Show that the Taylor series at the point zero defines a homomorphism

$\mathscr{E}_n \to \mathbb{R}\,[[x_1, \ldots, x_n]]$ into the ring of formal power series in n variables. The kernel of this homomorphism is $\mathfrak{m}_n^\infty := \cap_{k=1}^\infty \mathfrak{m}_n^k$ (see 5).

7 Following the notation of 4: $\mathscr{E}_n/\mathfrak{m}_n \cong \mathbb{R}$; consequently $\mathfrak{m}_n/\mathfrak{m}_n^2 \cong \mathbb{R}^n$. Show that a germ $\bar{f}: (\mathbb{R}^n, 0) \to (\mathbb{R}^m, 0)$ induces $f^*: \mathscr{E}_m \to \mathscr{E}_n$, $f^*\mathfrak{m}_m \subset \mathfrak{m}_n$, and so one obtains a linear map

$$f^*: \mathbb{R}^m \cong \frac{\mathfrak{m}_m}{\mathfrak{m}_m^2} \to \frac{\mathfrak{m}_n}{\mathfrak{m}_n^2} \cong \mathbb{R}^n.$$

This is given by the matrix ${}^t Df_0$.

8 Show that if the map $f: S^n \to \mathbb{R}$ is differentiable, then there are two different points $p, q \in S^n$, so that $T_p(f)$ and $T_q(f)$ are both 0.

9 Let $M = \{x \in \mathbb{R}^n \mid x_1^2 = x_2^2 + x_3^2 + \ldots + x_n^2 \text{ and } x_1 \geqslant 0\}, n > 1$. Show that M is *not* a differentiable submanifold of $\mathbb{R}^n$.

10 Let $f: \mathbb{R}^n \to \mathbb{R}^k$ be a differentiable map, so that for every real number t one has $f(t \cdot x) = t \cdot f(x)$. Show that f is linear.

11 Let $f: \mathbb{R}^n \to \mathbb{R}^k$, $f(0) = 0$ be a differentiable map, and let $f_t(x) = t^{-1}f(tx)$. Show that $f_t(x)$, if extended to $t = 0$ by Df_0, depends differentiably on t, x.

12 Let $f: \mathbb{R} \to \mathbb{R}$ be a differentiable function for which $f(0) = f'(0) = \ldots = f^{(n-1)}(0) = 0$, and $f^{(n)}(0) > 0$. Show that there exists an invertible germ $\bar{h}: (\mathbb{R}, 0) \to (\mathbb{R}, 0)$, so that $\bar{f} \cdot \bar{h} = \bar{x}^n$.

3
Vector bundles

Through the construction of tangent spaces there is a vector space associated
to every point of a manifold. In general, in differential topology and in
topology, there is often occasion to attach a vector space to every point of
a manifold or of a topological space, so that one has not just one single vector
space, but rather a whole 'bundle' of vector spaces, as in Fig. 16.

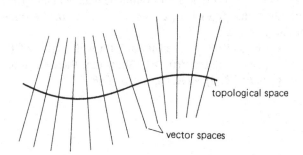

topological space

vector spaces

Fig. 16

(3.1) **Definition.** A (n-dimensional real topological) *vector bundle*
is a triple (E, π, X), where $\pi: E \to X$ is a continuous surjective map, every
$E_x := \pi^{-1}(x)$ has the structure of an n-dimensional real vector space such that:

Axiom of local triviality. Every point of X has a neighbourhood U,
for which there exists a homeomorphism

$$f: \pi^{-1}(U) \to U \times \mathbb{R}^n$$

such that for every $x \in U$

$$f_x := f|E_x: E_x \to \{x\} \times \mathbb{R}^n$$

is a vector space isomorphism, see Fig. 17.

Notation. (E, π, X) is called a vector bundle *over* X; E is called the
total space; E_x the *fibre*; X the *base*; and π the *projection* of the bundle.
Instead of (E, π, X) one usually writes E for short.

22

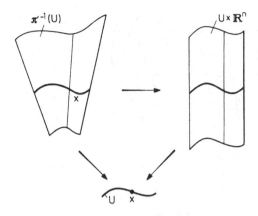

Fig. 17

(3.2) Definition. (f, U) as in the axiom of local triviality is called a *bundle chart*. A bundle over X is called *trivial* if it has a bundle chart (f, X).

The vector bundles over a fixed space X form in a natural manner the objects of a category. The corresponding 'morphisms' are the so-called 'bundle homomorphisms', which we shall now define.

(3.3) Definition. E and E' are vector bundles over X. A continuous map $f \colon E \to E'$ is called a *bundle homomorphism* if

$$E \xrightarrow{f} E'$$

$$\pi \searrow \quad \swarrow \pi'$$

$$X$$

is commutative and every $f_x \colon E_x \to E'_x$ is linear.

(3.4) Definition. If E is an n-dimensional vector bundle over X and $E' \subset E$ is a subset, so that around every point in X there is a bundle chart (f, U) with

$$f(\pi^{-1}(U) \cap E') = U \times \mathbb{R}^k \subset U \times \mathbb{R}^n,$$

then $(E', \pi|E', X)$ is in a natural manner a vector bundle over X and is called a k-dimensional *subbundle* of E, see Fig. 18.

For example if $f \colon E \to F$ is a bundle homomorphism of constant rank $rkf_x = \text{const}$, then

$$\text{kernel } f := \bigcup_{x \in X} \text{kernel } f_x$$

is a subbundle of E and

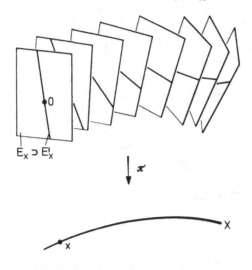

Fig. 18

$$\text{image } f := \bigcup_{x \in X} \text{image } f_x$$

is a subbundle of F. This is obvious because we have

(3.5) Rank theorem for bundle homomorphisms. Let $f: E \dashrightarrow F$ be a bundle homomorphism of constant rank $rk f_x = k$, of vector bundles over X, then around every point $x \in X$ there are bundle charts (ϕ, U) for E and (ψ, U) for F, such that for every $u \in U$ one has $(\psi \circ f \circ \phi^{-1})_u(v^1, \dots, v^m) = (v^1, \dots, v^k, 0, \dots, 0)$.

$$
\begin{array}{ccc}
E|U & \xrightarrow{\ \ f\ \ } & F|U \\
\phi \downarrow & & \downarrow \psi \\
U \times \mathbb{R}^m & \longrightarrow & U \times \mathbb{R}^n, \ (u, (v^1, \dots, v^m)) \mapsto (u, (v^1, \dots, v^k, \\
& & 0, \dots, 0)).
\end{array}
$$

Proof. First we may look at f on arbitrary charts and therefore w.l.o.g. suppose that f is a bundle homomorphism $U \times \mathbb{R}^m \to U \times \mathbb{R}^n$, $(u, v) \mapsto (u, f_u(v))$. Here $f_u = (f_u^1, \dots, f_u^n): \mathbb{R}^m \to \mathbb{R}^n$ is a linear map of rank k, which may be described by a matrix (depending on u), and w.l.o.g. (after a suitable permutation of coordinates in $\mathbb{R}^m$ and $\mathbb{R}^n$) the submatrix of the first k rows and columns of the particular matrix f_x is non singular. But then the bundle homomorphism

$$\phi: U \times \mathbb{R}^m \to U \times \mathbb{R}^m, \ \phi_u(v) = (f_u^1(v), \dots, f_u^k(v), v^{k+1}, \dots, v^m)$$

is isomorphic on the fibre over $u = x$, and therefore w.l.o.g. isomorphic on every fibre (if the determinant of ϕ_u does not vanish at the point $u = x$, it does not vanish at nearby points either). Using this bundle homomorphism as a new chart for $U \times \mathbb{R}^m$, we must look at $f \cdot \phi^{-1}$ and obtain

$$(f \cdot \phi^{-1})_u : v \mapsto (v^1, \ldots, v^k, g_u^{k+1}(v), \ldots, g_u^n(v)).$$

Since this still has rank k, the last $n-k$ components $g^{k+1}, \ldots, g^n$ (for given $u \in U$) in fact only depend on the first k components $(v^1, \ldots, v^k)$ of v; in matrix notation:

$$
(f \cdot \phi^{-1})_u \;=\;
\left[
\begin{array}{c|c}
\begin{smallmatrix} 1 & & \\ & \ddots & \\ & & 1 \end{smallmatrix} & 0 \\
\hline
g & 0
\end{array}
\right]
$$

So we may also write

$$(f \cdot \phi^{-1})_u : v \mapsto (v^1, \ldots, v^k, g_u^{k+1}(v^1, \ldots, v^k), \ldots, g_u^n(v^1, \ldots, v^k)).$$

But then on the other side we have the chart $\psi : U \times \mathbb{R}^n \to U \times \mathbb{R}^n$, $\psi_u(w) = (w^1, \ldots, w^k, w^{k+1} - g_u^{k+1}(w^1, \ldots, w^k), \ldots, w^n - g_u^n(w^1, \ldots, w^k))$, and $(\psi \cdot f \cdot \phi^{-1})_u(v) = (v^1, \ldots, v^k, 0, \ldots, 0)$. $\qquad\square$

Having thus refreshed ourselves in the oasis of a proof, we now turn again into the desert of definitions. First, we must mention another viewpoint from which we may consider bundles as being contained within other bundles.

(3.6) Definition. If (E, π, X) is a vector bundle and $X_0 \subset X$, then $(\pi^{-1}(X_0), \pi | \pi^{-1}(X_0), X_0)$ is a vector bundle over X_0, which is usually written as $E | X_0$ and is called the *restriction* of E to X_0, see Fig. 19.

(3.7) Definition ('section'). By a *section* of a vector bundle (E, π, X) we mean a continuous map $\sigma : X \to E$ with $\sigma(x) \in E_x$ for all $x \in X$. For example, every vector bundle has a 'zero-section'

$$X \to E$$

$$x \mapsto 0 \in E_x, \text{ see Fig. 20.}$$

(3.8) Note that if $\sigma : X \to E$ is a section, then $\sigma : X \to \sigma(X)$ is a homeomorphism.

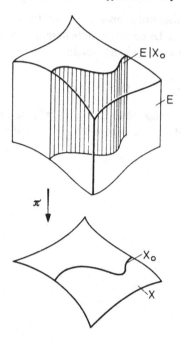

Fig. 19

$$X \to E$$
$$x \to 0 \in E_x$$

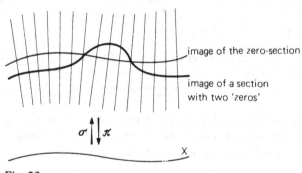

image of the zero-section

image of a section
with two 'zeros'

Fig. 20

In particular therefore, one can without harm identify the image of the zero-section with the base space X itself since via the zero-section one has a canonical homeomorphism.

From one vector bundle one can 'induce' new vector bundles. Suppose we are given an n-dimensional vector bundle E over Y and a continuous map $f: X \to Y$:

Thus we obtain the induced bundle f^*E over X by attaching the fibre $E_{f(x)}$ to every $x \in X$. This may be described as:

(3.9) Definition. Let (E, π, Y) be a vector bundle over Y and $f: X \to Y$ be continuous. Let us consider the graph of f and the canonical homeomorphism graph $(f) \cong X$, see Fig. 21. Then by the composition

$$f^*E := (X \times E) \mid \mathrm{Graph}\,(f) \subset X \times E$$
$$f^*\pi \left\downarrow \begin{array}{c} \downarrow \\ \mathrm{Graph}\,(f) \subset X \times Y \\ \downarrow \\ X \end{array}\right.$$

we define a vector bundle $(f^*E, f^*\pi, X)$, which is called the *bundle induced by f*.

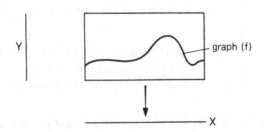

Fig. 21

(3.10) Note that the total space of f^*E is $\{(x, e) \mid \pi(e) = f(x)\} \subset X \times E$. This space is also called the *fibre product* of f and π.

The map $f^*E \to E$ given by the projection $X \times E \to E$ maps every fibre of f^*E linearly and isomorphically to a fibre of E. Such maps are called *bundle maps*. As a notion, including bundle homomorphisms and bundle maps as special cases, one also considers the quite general *linear maps* which are only required to map fibres linearly to fibres:

(3.11) Definition. If E, F respectively, are vector bundles over X, Y respectively, and $f: X \to Y$ is continuous, then a continuous map $\tilde{f}: E \to F$ is called a *linear map* over f if $\tilde{f}$ maps every fibre E_x linearly into $F_{f(x)}$:

$$E \xrightarrow{\tilde{f}} F$$
$$\downarrow \quad \downarrow$$
$$X \xrightarrow{f} Y \ ,$$

If these maps are isomorphisms $E_x \cong F_{f(x)}$ as well, then $\tilde{f}$ is called a *bundle map* over f.

The reason for explaining, just here, the terminology of bundle homomorphisms, bundle maps, and linear maps is that the construction of the induced bundle shows how one can write every linear map as the composition of a bundle homomorphism and a bundle map:

(3.12) Note that if $\phi \colon E \to F$ is a linear map of vector bundles over f and if $\tilde{f} \colon f^*F \to F$ is the canonical bundle map, then there is one and only one bundle homomorphism $h \colon E \to f^*F$ so that $\phi = \tilde{f} \circ h$:

$$E \xrightarrow{h} f^*F \xrightarrow{\tilde{f}} F$$
$$\pi \searrow \downarrow \qquad \downarrow$$
$$X \xrightarrow{f} Y \ ,$$

namely, $h(v) = (\pi(v), \phi(v)) \in X \times E$. This is called the *universal property* of the induced bundle.

Up to now we have only considered 'topological' vector bundles. We now wish to introduce the concept of differentiable vector bundle over a differentiable manifold. In order to do this we must first discuss the concept of the bundle atlas.

(3.13) Definition. Let (E, π, X) be an n-dimensional vector bundle. A set $\{(f_\alpha, U_\alpha) | \alpha \in A\}$ of bundle charts is called a *bundle atlas* for E, if $\cup_{\alpha \in A} U_\alpha = X$. The continuous mappings given by overlapping of the bundle charts

$$U_\alpha \cap U_\beta \to GL(n, \mathbb{R})$$

$$x \mapsto f_{\beta x} \circ f_{\alpha x}^{-1}$$

are called the *transition functions* of the atlas, see Fig. 22.

(3.14) Definition. A bundle atlas for a vector bundle over a differentiable manifold is differentiable if all its transition functions are differentiable. A differentiable vector bundle is a pair $(E, \mathfrak{B})$ consisting of a vector bundle E over M and a maximal differentiable bundle atlas $\mathfrak{B}$ for E.

(3.15) Note that the total space of a k-dimensional differentiable vector bundle over an n-dimensional manifold M is naturally an $(n + k)$-dimensional differentiable manifold.

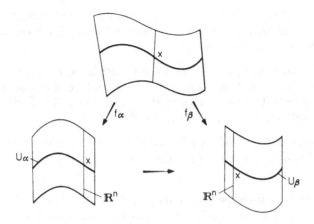

Fig. 22

Remark. The definitions and statements made up to now about topological vector bundles apply in an obvious manner to differentiable vector bundles.

We often encounter differentiable and topological vector bundles in a form in which one could perhaps call them 'pre-vector bundles'. Given are the usual defining terms

E total space

π projection

X base

$\mathfrak{B}$ bundle atlas

with the sole omission that the topology on E is not yet defined. E appears, for the present, simply as the union of the (disjoint!) vector spaces $E_x = \pi^{-1}(x)$. However, one can construct this topology in a canonical way and thereby obtain a real vector bundle.

As we obtain very many of our geometrically relevant vector bundles by these means, we shall make the notion of a 'pre-vector bundle' more precise:

(3.16) Definition. An n-dimensional *pre-vector bundle* is a quadruple $(E, \pi, X, \mathfrak{B})$ consisting of a set (!) E, a topological space X, a surjective mapping $\pi: E \to X$ with a vector space structure on every $E_x := \pi^{-1}(x)$, and a 'pre-bundle atlas' $\mathfrak{B}$, that is, a set $\{(f_\alpha, U_\alpha) | \alpha \in A\}$, where $\{U_\alpha | \alpha \in A\}$ is an open covering of X and

$$f_\alpha: \pi^{-1}(U_\alpha) \to U_\alpha \times \mathbb{R}^n$$

a bijective map which maps the fibre E_x linearly and isomorphically onto $\{x\} \times \mathbb{R}^n$ for every $x \in U_\alpha$ in such a way that all the transition functions $U_\alpha \cap U_\beta \to GL(n, \mathbb{R})$ of $\mathfrak{B}$ are continuous.

(3.17) First, note that if $(E, \pi, X, \mathfrak{B})$ is a pre-vector bundle, then there is exactly one topology on E, relative to which (E, π, X) is a vector bundle and $\mathfrak{B}$ is a bundle atlas thereof.

(3.18) Second, note that if M is a differentiable manifold and $(E, \pi, M, \mathfrak{B})$ is differentiable pre-vector bundle, that is, if all the transition functions of $\mathfrak{B}$ are differentiable, then by the maximal extension $\hat{\mathfrak{B}}$ of $\mathfrak{B}$ we clearly have a differentiable vector bundle $(E, \hat{\mathfrak{B}})$ over M.

Our first application, for whose sake alone the whole sequence of definitions would have been worthwhile, is the construction of the tangent bundle.

(3.19) Definition (tangent bundle). Let M be a differentiable n-dimensional manifold and $\mathfrak{A}$ be a differentiable atlas of M. Then we are given a differentiable pre-vector bundle $(TM, \pi, M, \mathfrak{B})$ as follows:

$$TM := \cup_{p \in M} T_p M$$

π: canonically $(T_p M \to p)$

where

$$\mathfrak{B} := \{(f_h | (h, U) \in \mathfrak{A}\},$$

$$f_h: \pi^{-1}(U) \to U \times \mathbb{R}^n$$

$$X \mapsto p \times (v_1, \ldots, v_n)$$

is given by the 'physical' coordinates $v_i = X(h_i)$ of $X \in T_p M$ with respect to (h, U), see (2.5) and Fig. 23.

The differentiable n-dimensional vector bundle TM over M given in this way, which is clearly independent of the choice of atlas, is called the *tangent bundle* of M.

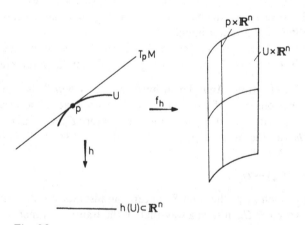

Fig. 23

(3.20) Definition. Let M be a differentiable manifold. By a (differentiable) *vector field* on M one understands a (differentiable) section

$$M \to TM$$

of the tangent bundle.

(3.21) Definition. If $f: M \to N$ is a differentiable map, then the differentials

$$T_p f: T_p M \to T_{f(p)} N$$

defines a differentiable map

$$Tf: TM \to TN$$

(as one can see from (2.4)), which is called *the differential* of f.

(3.22) Note that the differential is a 'linear map of vector bundles'. As remarked earlier (3.12), there is one and only one bundle homomorphism $TM \to f^* TN$, so that the diagram

$$TM \xrightarrow{Tf} TN$$

$$f^* TN$$

commutes.

(3.23) Exercises

1 Let U be a topological space and $f: U \to M(n \times k, \mathbb{R})$ a mapping into the space of real $(n \times k)$-matrices. Show that the map given by f

$$F: U \times \mathbb{R}^k \to \mathbb{R}^n$$

$$(u, x) \mapsto f(u) \cdot x$$

is continuous if and only if f is continuous. Show also that if U is a manifold then F is differentiable if and only if f is differentiable.

Remark. We have already made implicit use of this statement in the text.

2 Let (E, π, X) be a vector bundle over a connected space X, let $f: E \to E$ by a bundle homomorphism and $f \circ f = f$. Show that f has constant rank.

3 Let (E, π, X) be a vector bundle over a connected space X and $f: E \to E$ a bundle homomorphism with $f \circ f = \mathrm{Id}_E$. Show that Fix $(f) := \{v \in E \mid f(v) = v\}$ is a subbundle of E.

4 Let E be a vector bundle over X, let $X_0 \subset X$ be a subspace and $i: X_0 \subset X$ the inclusion. Show that $i^* E$ and $E | X_0$ are naturally isomorphic.

5 Show that if (E, π, X) is a trivial vector bundle, then every induced bundle f^*E (for $f: Y \to X$) is also trivial.

6 Let (E, π, X) be a vector bundle and $\pi_0 := \pi \,|\, E - \{$zero-section$\}$. Construct a nowhere vanishing 'canonical' section of π_0^*E.

7 Show that a vector bundle is trivial if and only if it possesses a bundle atlas, all of whose transition functions are maps into $\{$Id$\} \subset GL(n, \mathbb{R})$.

8 Over $\mathbb{R}\mathbf{P}^n = S^n/\sim$ let us consider the 1-dimensional subbundle

$$\eta_n := \{([x], \lambda x) \,|\, x \in S^n, \lambda \in \mathbb{R}\}$$

of $\mathbb{R}\mathbf{P}^n \times \mathbb{R}^{n+1}$. (Why is it a subbundle?) Prove that for $n \geqslant 1$, η_n is non-trivial.

Hint: consider $\eta_n - \{$zero-section$\}$.

9 Prove that every 1-dimensional vector bundle over S^1 is either trivial or isomorphic to the bundle

$$\begin{array}{c} \eta_1 \\ \downarrow \\ S^1 \cong \mathbb{R}\,\mathbf{P}^1 \end{array}$$

The surface η_1 is also called the (unbounded) Möbius band (see Fig. 6 and Fig. 24).

10 Prove: if one removes a point from $\mathbb{R}\mathbf{P}^{n+1}$, then one obtains a manifold which is diffeomorphic to the total space of η_n:

$$\mathbb{R}\mathbf{P}^{n+1} - pt \cong \eta_n$$

Hint: w.l.o.g. $pt = [0, \ldots, 0, 1]$.

11 Let $n \geqslant 1$. Show that there exist precisely two isomorphism types of n-dimensional vector bundles over S^1 (see exercise 9).

12 Show that $TS^1 \cong S^1 \times \mathbb{R}$.

13 Show that the tangent bundle of S^2 possesses an atlas with two bundle charts.

14 Let M be connected. Show that a differentiable map $f: M \to N$, whose differential Tf is everywhere zero, must be constant.

15 Show that if $f: M \to N$ is an embedding, then so is $Tf: TM \to TN$.

16 Construct a vector field on S^2 which has exactly two zero points.

17 Construct a vector field on S^2 which has exactly one zero point.

18 Let $M \subset \mathbb{R}^n$ be a submanifold. Show that

$$TM \cong \{(x, v) \in M \times \mathbb{R}^n \,|\, v \in T_xM \subset \mathbb{R}^n\}.$$

19 Show that the submanifold of $\mathbb{C}^{n+1}$

$$E = \{(z_0, \ldots, z_n) \in \mathbb{C}^{n+1} \,|\, z_0^2 + \ldots + z_n^2 = 1\}$$

is diffeomorphic to the total space of the tangent bundle of the unit sphere S^n.

4
Linear algebra for vector bundles

The algebraic operations which one employs in linear algebra for vector spaces and homomorphisms can usually also be given a meaning on vector bundles and bundle homomorphisms, by operating in the fibres at every point of the base, as one has learned to do in linear algebra. For example, one constructs the direct sum $E \oplus F$ (the so-called 'Whitney sum') of two vector bundles E and F over X, by using the direct sum $E_x \oplus F_x$ as fibre of $E \oplus F$ at every point $x \in X$, etc.

Of course, we must explain more precisely the bundle structure of the families of vector spaces, which arise in this way.

(4.1) Definition as typical example. Let E and F be vector bundles over X with bundle atlases $\mathfrak{A}$ and $\mathfrak{B}$. Then a pre-vector bundle $E \oplus F$ is given in the following manner:

$$E \oplus F := \bigcup_{x \in X} E_x \oplus F_x$$

projection: canonical

atlas: $\{\phi \oplus \psi, U \cap V | (\phi, U) \in \mathfrak{A}, \ (\psi, V) \in \mathfrak{B}\}$

where $\phi \oplus \psi$ is to be understood in the following way:

$$E_x \oplus F_x \xrightarrow{\phi_x \oplus \psi_x} \{x\} \times \mathbb{R}^n \oplus \mathbb{R}^k.$$

The vector bundle $E \oplus F$ associated with this pre-vector bundle is called the *Whitney sum* of E and F.

(4.2) Supplement. If $f: E \to E'$ and $g: F \to F'$ are bundle homomorphisms, then a bundle homomorphism $f \oplus g: E \oplus F \to E' \oplus F'$ is defined in a canonical manner.

(4.3) Note that if E and F are differentiable, then in a natural manner so is $E \oplus F$; if f and g are differentiable, then so is $f \oplus g$.

(4.4) Further examples. Analogously, one transfers other notions of linear algebra 'fibrewise' to the category of topological, respectively differentiable vector bundles over X. Thus, for example, one obtains:

(i) *tensor product $E \otimes F$,*

(ii) *quotient bundle E/F (when F is a subbundle of E),*

(iii) *dual bundle E^*,*

(iv) *homomorphism bundle* Hom (E, F),

(v) *bundle $Alt^k(E)$ of alternating k-forms,*

(vi) *bundle $\Lambda^k E$ of k-fold exterior powers,*

for vector bundles E, F over X and, in a natural manner also, the relevant bundle homomorphisms.

Remark. One must note carefully that some of the functors of linear algebra, which have here been carried over to bundles, are *contravariant*. For example, Hom in the first variable: bundle homomorphisms $f: A \to B$ and $g: F \to F'$ induce a bundle homomorphism

$$\text{Hom} (f, g): \text{Hom} (B, F) \to \text{Hom} (A, F'),$$

namely, by

$$
\begin{array}{ccc}
B & \to & F \\
f \uparrow & & \downarrow g \\
A & \dashrightarrow & F'.
\end{array}
$$

Correspondingly, the bundle charts of Hom (E, F) are obtained from bundle charts (ϕ, U) of E and (ψ, V) of f in the form

$$\text{Hom} (\phi^{-1}, \psi): \text{Hom} (E, F)|U \cap V \to (U \cap V) \times \text{Hom} (\mathbb{R}^n, \mathbb{R}^k)$$
$$= (U \cap V) \times \mathbb{R}^{nk}.$$

The term orientation requires careful consideration. Naturally, one orients a vector bundle by orienting each fibre, and in such a way that, for an arbitrary continuous path in the base, the orientation does not suddenly 'jump'.

(4.5) Definition (orientation of a vector bundle). Let E be an n-dimensional vector bundle over X. A family

$$\mathfrak{o} = \{\mathfrak{o}_x\}_{x \in X}$$

of orientations $\mathfrak{o}_x$ of the fibres E_x is called an *orientation* of E if about every point of X there is a bundle chart (f, U) for E, so that by means of $f_u: E_u \cong \mathbb{R}^n$ the orientation $\mathfrak{o}_u$ for every $u \in U$ is transferred to the same fixed orientation of $\mathbb{R}^n$.

Whereas up to now we have been able to transfer the constructions of linear algebra simply by way of fibres or by way of charts to vector bundles, we now for the first time come upon a global phenomenon: for *one* vector space, *one* fibre, we can always choose an orientation, but the whole bundle need not be orientable. If one were to orient a specific fibre E_x, then this orientation simply extends through the charts (f, U) in (4.5) to the fibres over points in the neighbourhood U of x.

If, however, one attempts to orient all of E in this way, by passing from chart to chart, then one notices with certain bundles that, at some point, this procedure has to lead to a jump in the orientation, as illustrated by Fig. 24. However, one must at times also pose questions of orientation for non-orientable bundles (for example in the proof of non-orientability or whether one may apply a certain known theorem for orientable bundles to non-orientable bundles also), and for this it is very useful to employ the concept of orientation cover, which is defined for every bundle.

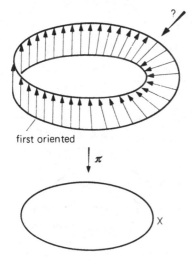

first oriented

Fig. 24

(4.6) Definition and notation. Let (E, π, X) be an n-dimensional vector bundle and $\Lambda^n E$ the 1-dimensional nth exterior power bundle. If one defines an equivalence relation in $\Lambda^n E - \{\text{zero-section}\}$ by $x \sim y \Leftrightarrow y = \lambda x$ for some $\lambda > 0$ and introduces the quotient topology on the set $\tilde{X}(E)$ of equivalence classes, then the canonical projection

$$\begin{array}{c} \tilde{X}(E) \\ \downarrow \tilde{\pi} \\ X \end{array}$$

is a two leaved covering of X and is called the *orientation cover* of E, see Fig. 25.

The relation known from linear algebra between orientation and n-fold exterior product (two bases $(v_1, \ldots, v_n)$ and $(w_1, \ldots, w_n)$ have the same

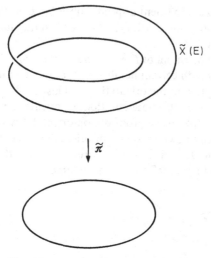

$\tilde{X}$ (E)

$\downarrow \tilde{\pi}$

Fig. 25

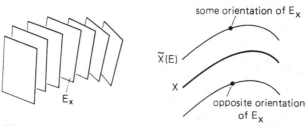

some orientation of E_x

$\tilde{X}(E)$

X

opposite orientation
of E_x

E_x

Fig. 26

orientation if and only if $v_1 \wedge \ldots \wedge v_n$ and $w_1 \wedge \ldots \wedge w_n$ differ only by some positive factor) shows immediately that $\tilde{X}(E)$, as a set, is canonically the same as the set of all orientations of all fibres, and $\tilde{\pi}^{-1}(x)$ consists of the two orientations of E_x, see Fig. 26. One may also think of $\tilde{X}(E)$ in this way; the description as

$$(\Lambda^n E - \text{zero-section})/\sim$$

has the technical advantage of immediately giving the topology on $\tilde{X}(E)$.

(4.7) *Note* that E is orientable if and only if $\tilde{X}(E)$ is a trivial cover, that is, isomorphic to $X \times \mathbb{Z}_2$. An orientation of E is then to be thought of as a section $X \xrightarrow{\mathrm{o}} \tilde{X}(E)$ (continuous mapping with $\pi \cdot \mathrm{o} = \mathrm{Id}_X$).

(4.8) *Note* that the cover $\tilde{X}(E)$ is also canonically isomorphic (that is, could also have been described as) $(\Lambda^n E^* - \text{zero-section})/\sim$ and to $(Alt^n E - \text{zero-section})/\sim$.

(4.9) Definition (orientation of a manifold). By an *orientation* of a manifold M, one means an orientation of the tangent bundle TM.

Another concept taken from linear algebra, the carrying over of which to vector bundles requires some attention, is that of scalar product.

If V is a real vector space then, as is known, one can consider the bilinear forms

$$V \times V \to \mathbb{R}$$

as the elements of $(V \otimes V)^*$. If E is a vector bundle over X then, by (4.4), the bundle $(E \otimes E)^*$ is defined and we come to:

(4.10) Definition (scalar product, Riemannian metric). If (E, π, X) is a vector bundle then, by a *scalar product* or a *Riemannian metric* for E, we mean a continuous section

$$s : X \to (E \otimes E)^*$$

such that for every $x \in X$ the bilinear form determined by this

$$E_x \times E_x \to \mathbb{R}$$

$$(v, w) \mapsto \langle v, w \rangle_x$$

is symmetric and positive definite. The metric is *differentiable* if X is a manifold and E and s are differentiable.

(4.11) Remark. If the vector bundle E is equipped with a Riemannian metric and $F \subset E$ is a subvector bundle, then

$$F^{\perp} := \bigcup_{x \in X} F_x^{\perp}$$

is also a subvector bundle.

Proof. If (f, U) is a bundle chart of E, which represents $F|U$ as $U \times (\mathbb{R}^k \times 0) \subset U \times \mathbb{R}^n$, and if $v_1, \ldots, v_n$ are sections of $E|U$, which under f correspond to the canonical basis vectors of $\mathbb{R}^n$, then one obtains by means of the Schmidt orthogonalisation process sections $v_1', \ldots, v_n'$ of $E|U$, which form an orthonormal basis of E_x for every $x \in U$, and in such a way that $v_1'(x), \ldots, v_k'(x)$ precisely span F_x, and $v_{k+1}'(x), \ldots, v_n'(x)$ span $F_x^{\perp}$.
Therefore,

$$f' : E|U \to U \times \mathbb{R}^n$$

$$\lambda_1 v_1'(x) + \ldots + \lambda_n v_n'(x) \mapsto (x, \lambda_1, \ldots, \lambda_n)$$

defines a bundle chart, which represents $F|U$ as $U \times \mathbb{R}^k$ and $F^{\perp}|U$ as the complementary $U \times \mathbb{R}^{n-k}$. $\qquad\square$

Since f' is obviously orthogonal in every fibre, we can note the following as a subsiduary of the proof:

(4.12) Note that every vector bundle with a Riemannian metric possesses a bundle atlas consisting of fibrewise orthogonal bundle charts. In particular, the transition functions of such an atlas are maps into $O(n) \subset GL(n, \mathbb{R})$.

(4.13) Note that if E is equipped with a Riemannian metric and $F \subset E$ is a subbundle, then the composition

$$F^\perp \subset E \xrightarrow[\text{Proj.}]{} E/F$$

is a bundle isomorphism $F^\perp \cong E/F$; one can thus consider E/F simply as $F^\perp$.

For dimensional reasons one has only to consider that the kernel of this composition vanishes. For every fibre this means $F_x^\perp \cap F_x = 0$.

This clear presentation of the quotient bundle as an orthogonal complement should, in particular, be kept in mind when considering the normal bundle of a submanifold.

(4.14) Definition (normal bundle). If M is a differentiable manifold and $X \subset M$ is a submanifold, then 'normal X'

$$\perp X := (TM \mid X)/TX$$

is called the *normal bundle* of X in M, see Fig. 27.

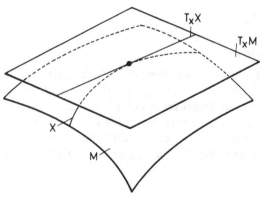

Fig. 27

(4.15) Definition (Riemannian manifold). A manifold M, whose tangent bundle has a differentiable scalar product, is called a *Riemannian manifold* ('a Riemannian manifold is a pair $(M, <, >)$, consisting ... ').

(4.16) Note that if M is a Riemannian manifold and $X \subset M$ a submanifold, then the normal bundle of X in M is canonically isomorphic to $(TX)^\perp$, see Fig. 28.

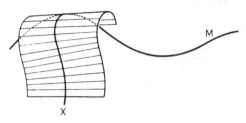

Fig. 28

Now we come to the important question of the existence of Riemannian metrics on vector bundles. Let (E, π, X) be a vector bundle. We look for a section

$$s: X \to (E \otimes E)^*,$$

so that every $s(x)$ is symmetric and positive definite. It is, of course, quite easy to find such a section for $E|U$ for every bundle chart (f, U) of E, we need only to begin with the usual scalar product in $\mathbb{R}^n$; $E|U \cong U \times \mathbb{R}^n$. If we do this for every chart of a bundle atlas, then we come to the following situation: We have 'local' sections (illustrated in Fig. 29):

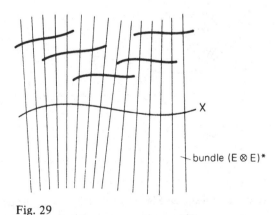

Fig. 29

However, we are looking for a global section (illustrated in Fig. 30):

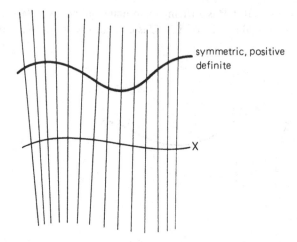

symmetric, positive
definite

X

Fig. 30

One often faces such a problem in topology, and it can be quite difficult or
insoluble (orientation!). However, there is help at hand if the property
required of the vectors $s(x)$ is a 'convex' property, that is if, with $s_1(x)$ and
$s_2(x)$ also, all

$$(1 - t)s_1(x) + ts_2(x)$$

have this property for $t \in [0, 1]$:

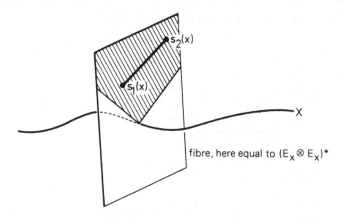

$s_2(x)$

$s_1(x)$

X

fibre, here equal to $(E_x \otimes E_x)^*$

Fig. 31

Symmetry and positive definiteness (see Fig. 31) are such convex properties.
 The technical tool, with which one stitches together locally given sections
to form a global section — something which the differential topologist must
always have at hand — is a partition of unity:

(4.17) Definition. Let X be a topological space. A family $\{\tau_\alpha\}_{\alpha \in A}$ of continuous functions

$$\tau_\alpha : X \to [0, 1]$$

is called a *partition of unity* if every point in X has a neighbourhood in which only finitely many of the τ_α are different from zero and for all $x \in X$ we have

$$\sum_{\alpha \in A} \tau_\alpha(x) = 1.$$

(4.18) Definition. Such a partition of unity is said to be *subordinate* to a given covering of X if for every α the *support* of τ_α (that is, Supp $\tau_\alpha :=$ $\{x \in X \,|\, \tau_\alpha(x) \neq 0\}$) is entirely contained in one of the covering subsets.

(4.19) Theorem quoted from general topology (see [8], p. 171; for manifolds see Chapter 7.): *If X is paracompact, then for every open cover there exists a subordinate partition of unity.*

(4.20) Corollary. *If E is a vector bundle over a paracompact space (e.g. a manifold), then one can equip E with a Riemannian metric.*

Proof. Let $\mathfrak{A}$ be an atlas for E and $\{\tau_\alpha\}_{\alpha \in A}$ a partition of unity subordinate to $\{U\}_{(f,\, U) \in \mathfrak{A}}$. For every α we choose a bundle chart (f_α, U_α), so that Supp $\tau_\alpha \subset U_\alpha$, and a Riemannian metric s_α for $E|U_\alpha$. Then $\tau_\alpha s_\alpha$ is a continuous section of $(E \otimes E)^*$ defined on all of X, if one understands $\tau_\alpha s_\alpha$ as being given by the zero section outside the support of τ_α, see Fig. 32.

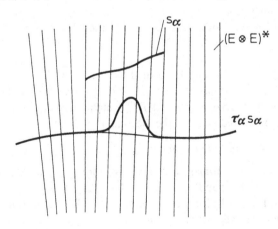

Fig. 32

Then $s := \sum_{\alpha \in A} \tau_\alpha s_\alpha$ is clearly a Riemannian metric for X. $\qquad\square$

(4.12) Remark. On differentiable manifolds there is even a differentiable subordinate partition of unity for every open covering, that is, the τ_α can be chosen to be differentiable, and consequently every differentiable

vector bundle also has a differentiable Riemannian metric. Because of the great importance of differentiable partitions of unity in differential topology, we do not wish to close our treatment of them with this remark. Their existence will be proved in detail in Chapter 7 and, until then, we shall refrain from making use of them.

(4.22) Exercises

1 Show how the bundle homomorphisms

$$f: E \to F$$

can be considered as sections in $E^* \otimes F = \text{Hom}\,(E, F)$.

2 Prove that if $E_1 \oplus E_2 \cong E_3$ and if two of the vector bundles E_i are orientable, then the third is orientable as well.

3 Let E be an orientable vector bundle and $F \subset E$ a subbundle. Show that E/F is orientable if and only if F is orientable.

4 Prove that a vector bundle is orientable if and only if it possesses a bundle atlas, all of whose transition functions are maps in
$GL^+(n, \mathbb{R}) := \{A \in GL(n, \mathbb{R}) \,|\, \det A > 0\}$.

5 Let E be a vector bundle. Show that $E \oplus E$ is orientable.

6 Let (E, π, X) be a vector bundle and $\tilde{\pi} \colon \tilde{X}(E) \to X$ its orientation cover. Show that $\tilde{\pi}^* E$ possesses a (canonical) orientation.

7 By the orientation cover $\tilde{M} \to M$ of a manifold M, one means the orientation cover of TM. Show that the manifold $\tilde{M}$ is orientable.

8 Show that $\mathbb{R}P^n$ is orientable for odd values of n and non-orientable for even values of n.

9 Show that for every submanifold $M \subset \mathbb{R}^n$ the Whitney sum

$$TM \oplus \bot M$$

of the tangent bundle and the normal bundle is trivial.

10 A vector bundle is *stably trivial* if its Whitney sum with a suitable trivial bundle is trivial. Show that TS^n is stably trivial.

11 Let M be a manifold and Δ_M the diagonal in $M \times M$:

$$\Delta_M := \{(x, x) \in M \times M \,|\, x \in M\}.$$

Show that Δ_M is a submanifold of $M \times M$, for which the tangent bundle and normal bundle are isomorphic: $T\Delta_M \cong \bot \Delta_M$.

12 Show that if (E, π, X) is a trivial bundle with a Riemannian metric, then there is a bundle isomorphism

$$E \cong X \times \mathbb{R}^n,$$

which is an isometry in every fibre.

13 Let E be a vector bundle over X and $\mathfrak{A}$ a bundle atlas for E, all of whose transition functions are maps into $O(n) \subset GL(n, \mathbb{R})$. Show that there is precisely one Riemannian metric $\langle\,,\rangle$ on E, such that all charts of $\mathfrak{A}$ are isometries on the fibres.

14 Let X be a space (e.g. a manifold), on which there is a partition of unity subordinate to every open covering. Show that for every 'line bundle' (that is, 1-dimensional vector bundle) L over X, $L \otimes L$ is trivial.

15 Show that a product of two non-empty differentiable manifolds $M \times N$ is orientable if and only if both factors M, N are orientable.

16 Let $T = \{(z, w) \in \mathbb{C} \times \mathbb{C} \mid |z| = |w| = 1\}$ be the torus, and $\tau : T \to T$ the involution $\tau(z, w) = (-z, \overline{w})$. Using (1.11) exercise 4, give M/τ the structure of a differentiable manifold. It is called the 'Klein bottle'. Is it orientable?

5

Local and tangential properties

For the local study of manifolds it is, above all, important to see whether a germ $\bar{f} : (M, p) \to (N, q)$ is invertible, that is, whether a mapping maps a neighbourhood of p diffeomorphically onto a neighbourhood of q. The functorial property shows that for such a germ the differential $T_p f : T_p M \to T_q N$ is an isomorphism and the differential calculus shows that this condition is sufficient.

(5.1) Inverse function theorem. *A differentiable germ is invertible if and only if its differential is an isomorphism.*

If we introduce charts $\bar{h} : (M, p) \to (\mathbb{R}^m, 0)$ and $\bar{k} : (N, q) \to (\mathbb{R}^n, 0)$, then $\bar{f}$ induces the germ

$$\bar{g} = \bar{k} \circ \bar{f} \circ \bar{h}^{-1} : (\mathbb{R}^m, 0) \to (\mathbb{R}^n, 0).$$

The differential is then a linear map $\mathbb{R}^m \to \mathbb{R}^n$, which, by (2.4), is described by the Jacobi matrix at the origin Dg_0. If this is invertible (the differential an isomorphism, in particular $m = n$), then some representative g of $\bar{g}$ is invertible in some neighbourhood, that is, $\bar{g}$ and hence also $\bar{f}$ are invertible (see Lang [2], chapter 17, section 3, pp. 349).

In a yet more general situation a germ is described by its differential:

(5.2) Definition. The *rank* of a differentiable map $f : M \to N$ at the point $p \in M$ (the rank of the germ $\bar{f} : (M, p) \to N$) is the number

$$rk_p f := rk T_p f.$$

(5.3) Remark. The rank of a map is lower semi-continuous. If $rk_p f = r$, then there is a neighbourhood U of p, so that $rk_q f \geqslant r$ for all $q \in U$.

Proof. After choosing charts, one must show that the rank of a Jacobi matrix Df cannot decrease locally around $p \in V \subset \mathbb{R}^m$. The components of this matrix describe a differentiable map:

$$Df : V \to \mathbb{R}^{m \cdot n}, \quad q \mapsto \left(\frac{\partial f_i}{\partial x_j} (q) \right).$$

Because $rk_p f = r$, there is a $(r \times r)$-submatrix of Df_p (w.l.o.g., consisting of the first r rows and columns), whose determinant does not vanish at the point p, that is, the map

$$V \to \mathbb{R}^{m \cdot n} \longrightarrow \mathbb{R}^{r \cdot r} \longrightarrow \mathbb{R}$$

$$p \mapsto Df_p \longmapsto \text{submatrix} \longmapsto \text{determinant}$$

neither vanishes at the point p, nor in a neighbourhood U of p; the rank cannot decrease there. □

Of course, arbitrarily close to p, the rank can be greater than $rk_p f$, example:

$$f: \mathbb{R} \to \mathbb{R}, \qquad x \mapsto x^2$$

has the differential $Df_x = 2x \neq 0$ for $x \neq 0$.

If a germ $\bar{f}: (M, p) \to (N, q)$ is described for suitable charts around p and q by a linear map, that is, if there is a linear map $g: \mathbb{R}^m \to \mathbb{R}^n$ and charts h, k, so that the following diagram commutes,

$$
\begin{array}{ccc}
(M,p) & \xrightarrow{\ \bar{f}\ } & (N,q) \\
\downarrow{\scriptstyle h} & & \downarrow{\scriptstyle k} \\
(\mathbb{R}^m,0) & \xrightarrow{\ \bar{g}\ } & (\mathbb{R}^n,0),
\end{array}
$$

then the differential T_g is given by the Jacobi matrix, and the Jacobi matrix Dg of the linear map $g: x \mapsto y$ with

$$y_i = \sum_j a_{ij} x_j$$

is $(\partial y_i / \partial x_j) = (a_{ij})$, therefore constant. Thus the rank of a representative f is locally constant, that is, the same as the rank of the matrix (a_{ij}). This condition on the rank is not only necessary, but — as we shall soon see — also sufficient for describing the germ $\bar{f}$ by the differential $T_p f = g$, subject to the choice of suitable charts.

By suitable choice of bases, a linear map of rank r can always be taken as

$$g: \mathbb{R}^m \to \mathbb{R}^n, \qquad (x_1, \ldots, x_m) \mapsto (x_1, \ldots, x_r, 0, \ldots, 0).$$

We wish to say that a germ has constant rank if it possesses a representative with constant rank.

(5.4) Rank theorem (Fig. 33). *If $\bar{f}: (M, p) \to (N, q)$ is a germ of constant rank r, then there are charts h around p and k around q, so that the germ $\bar{k} \circ \bar{f} \circ \bar{h}^{-1}: (\mathbb{R}^m, 0) \to (\mathbb{R}^n, 0)$ is represented by the map*

$$(x_1, \ldots, x_m) \mapsto (x_1, \ldots, x_r, 0, \ldots, 0).$$

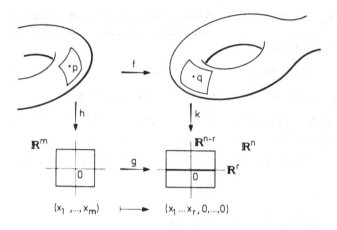

Fig. 33

Proof. We may immediately assume $\bar{f}\colon (\mathbb{R}^m, 0) \to (\mathbb{R}^n, 0)$; we then find an $(r \times r)$-submatrix of Df, which is regular at the origin, and after exchanging coordinates in $\mathbb{R}^m$ and $\mathbb{R}^n$ we obtain a matrix

$$\frac{\partial f_i}{\partial x_j}, \quad 1 \leqslant i, j \leqslant r$$

which is regular at the origin.

Let $\bar{h}\colon (\mathbb{R}^m, 0) \to (\mathbb{R}^m, 0)$ be represented by the map

$$h\colon (x_1, \ldots, x_m) \mapsto (f_1(x), \ldots, f_r(x), x_{r+1}, \ldots, x_m),$$

then the Jacobi matrix of h has the form

$$Dh = \begin{array}{c} \left.\begin{array}{|c|c|} \hline \multicolumn{2}{|c|}{\partial f_i/\partial x_j} \\ \hline & 1 \\ 0 & \ddots \\ & \ddots 1 \\ \hline \end{array}\right\} r \\ \hspace{3em} \left.\phantom{\begin{array}{c}x\\x\end{array}}\right\} m-r, \end{array}$$

$$\det(Dh_0) = \det(\partial f_i/\partial x_j(0))_{i,\,j \leqslant r} \neq 0.$$

Thus, by the inverse function theorem, $\bar{h}$ is an invertible germ and the diagram

$$(\mathbb{R}^m, 0) \xrightarrow{\ \bar{f}\ } (\mathbb{R}^n, 0) \qquad (x_1, \ldots, x_m) \mapsto (f_1(x), \ldots, f_n(x))$$

$$\bar{h} \searrow \qquad \swarrow \bar{g} = \bar{f} \cdot \bar{h}^{-1} \qquad \qquad \searrow \qquad \swarrow$$

$$(\mathbb{R}^m, 0) \qquad\qquad (f_1(x), \ldots, f_r(x), x_{r+1}, \ldots, x_m)$$

$$\parallel$$

$$(z_1, \ldots, z_m)$$

shows that the germ $\bar{g} = \bar{f} \circ \bar{h}^{-1}$ is represented by the map

(5.5) $(z_1, \ldots, z_m) \mapsto (z_1, \ldots, z_r, g_{r+1}(z), \ldots, g_n(z))$.

The Jacobi matrix of $\bar{g}$ therefore has the form

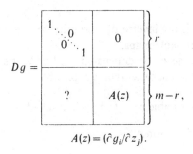

$$A(z) = (\partial g_i / \partial z_j).$$

Transformation in the pre-image space leads this far, and we have only made use of the fact that $rk_0 f \geqslant r$.

But now since $rk(f) = rk(g) = rk(Dg) = r$ in a neighbourhood of the origin, in this neighbourhood we must have $A(z) = 0$, therefore

(*) $\dfrac{\partial g_i}{\partial z_j} = 0$ for $r+1 \leqslant i \leqslant n, \ r+1 \leqslant j \leqslant m$.

Let the germ $\bar{k}: (\mathbb{R}^n, 0) \to (\mathbb{R}^n, 0)$ be represented in the image space by a map

$(y_1, \ldots, y_n) \mapsto (y_1, \ldots, y_r,$

$y_{r+1} - g_{r+1}(y_1, \ldots, y_r, 0, \ldots, 0), \ldots, y_n - g_n(y_1, \ldots, y_r, 0, \ldots, 0))$.

The Jacobi matrix of $\bar{k}$ has the form

and thus $\bar{k}$ is invertible and $\bar{k} \circ \bar{f} \circ \bar{h}^{-1} = \bar{k} \circ \bar{g}$ is represented by the composition

$(z_1, \ldots, z_m) \xmapsto{g} (z_1, \ldots, z_r, g_{r+1}(z), \ldots, g_n(z))$

$\xmapsto{k} (z_1, \ldots, z_r, g_{r+1}(z) - g_{r+1}(z_1, \ldots, z_r, 0, \ldots, 0), \ldots,$

$g_n(z) - g_n(z_1, \ldots, z_r, 0, \ldots, 0))$.

If we now restrict ourselves to a cube neighbourhood $|z_j| < \epsilon$ for sufficiently small ϵ, then

$$g_i(z_1, \ldots, z_n) - g_i(z_1, \ldots, z_r, 0, \ldots, 0) = 0, \quad r+1 \leqslant i \leqslant n$$

on account of $(*)$, thus $\bar{k} \cdot \bar{g}$ is represented by

$$(z_1, \ldots, z_m) \mapsto (z_1, \ldots, z_r, 0, \ldots, 0). \qquad \qquad \square$$

The rank theorem, in other words the inverse function theorem, dominates the elementary geometry of differentiable maps.

If $rk_p f$ is maximal, that is, the same as the dimension of M or N, then the rank is locally constant (5.3), and the rank theorem is applicable.

(5.6) Definition. A differentiable map $f: M \to N$ is called:

a *submersion*   if $rk_p f = \dim N$,

an *immersion*   if $rk_p f = \dim M$,

for all $p \in M$. A point $p \in M$ is *regular* if the differential $T_p f$ is surjective. A point $q \in N$ is a *regular value* of f if every point of $f^{-1}(q)$ is regular. Instead of 'non-regular' one can also say *singular* or *critical*.

Note in particular that a point $q \in N$ is a regular value, if $f^{-1}(q) = \emptyset$, that is, if it is not a value.

The map f is then a submersion if and only if every point $p \in M$ is regular, or every $q \in N$ is a regular value.

The statement that f is an immersion means that the differential Tf is injective at every point $p \in M$. Then by the rank theorem, locally in specific coordinates, $\bar{f}$ has the form

$$(x_1, \ldots, x_m) \mapsto (x_1, \ldots, x_m, 0, \ldots, 0).$$

In particular, every point of M possesses a neighbourhood which is embedded by f. However, f need not be injective, see Fig. 34, and even if f is injective, f need not be an embedding by definition (1.10). The obvious counter-example is illustrated in Fig. 35.

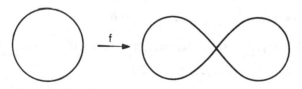

Fig. 34

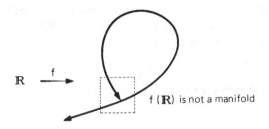

f (**R**) is not a manifold

Fig. 35

If, however, M is compact, $f: M \to N$ an immersion and injective, then f is an embedding; more generally:

(5.7) Theorem. *Let $f: M \to N$ be an injective immersion and $f: M \to f(M)$ be a homeomorphism where $f(M) \subset N$ carries the subspace topology. Then f is an embedding.*

Proof. If $p \in M$ and $f(p) = q \in N$, the rank theorem yields charts $h: U \to U' \subset \mathbb{R}^m$ and $k: V \to V' \subset \mathbb{R}^m \times \mathbb{R}^s$ around p and q, so that f induces the map

$$\tilde{f} = k \circ f \circ h^{-1}: x \mapsto (x, 0).$$

U is then chosen so small that f is defined on all of U', and $U' \times B \subset V'$ for some neighbourhood B of 0 in $\mathbb{R}^s$. Then let V be so shrunk that $U' \times B = V'$.

Since f is a homeomorphism, $U = f^{-1} W$ for some open neighbourhood W of q, and for the chart $k' := k|(V \cap W)$ we have $k'(f(M) \cap V \cap W) = \mathbb{R}^m \cap k'(V \cap W)$. Therefore, $f(M)$ is a submanifold of N and $f: M \to f(M)$ is both locally invertible and bijective and is therefore a diffeomorphism. Fig. 36 illustrates this proof. □

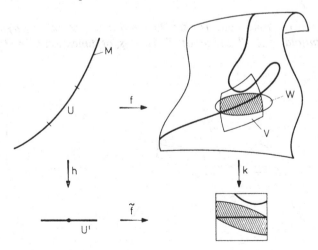

Fig. 36

For an immersion $f: M \rightarrow N$, as well as for an embedding, one can define a normal bundle. Since by definition (5.6), the map $Tf: TM \rightarrow TN$ maps every fibre injectively, the induced homomorphism (3.12)

$$h: TM \rightarrow f^*TN$$

of vector bundles over M is injective, and the quotient bundle

(5.8) $f^*TN/h(TM)$

is called the *normal bundle* of f.

(5.9) **Lemma.** *If q is a regular value of the differentiable map* $f: M^{n+k} \rightarrow N^n$, *then $f^{-1}(q)$ is a differentiable submanifold of M with co-dimension n.*

Proof. If $f(p) = q$, then by (5.3) the rank of f around p is locally constant because it cannot become larger than n. Therefore, using the rank theorem, one can introduce local coordinate systems around p and q so that, with respect to these coordinates in a neighbourhood U of p, f is given by

$$(x_1, \ldots, x_{n+k}) \mapsto (x_1, \ldots, x_n),$$

$$p = (0, \ldots, 0), \qquad q = (0, \ldots, 0).$$

Then $f^{-1}(q) \cap U = \mathbb{R}^k \cap U \subset \mathbb{R}^{n+k} \cap U$; thus $f^{-1}(q)$ is a submanifold of dimension k.   □

This lemma is the most important tool in showing that a subset of a differentiable manifold is a submanifold, or in constructing manifolds. For example, the contour lines of a (geographical) map are submanifolds, just so long as the height is regular, see Fig. 37. By way of illustration we give the following:

(5.10) **Application.** *The set $O(n)$ of real orthogonal $(n \times n)$-matrices is a submanifold of $\mathbb{R}^{n \cdot n}$, the set of all matrices, of dimension $\frac{1}{2} \cdot n \cdot (n-1)$.*

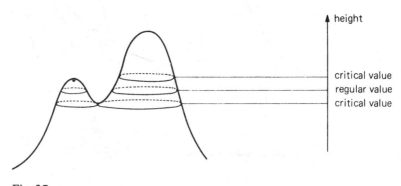

Fig. 37

Proof. A matrix $A \in \mathbb{R}^{n \cdot n}$ is orthogonal if and only if ^tAA is the identity matrix E. In any case, ^tAA is symmetric. Therefore, $O(n)$ is the pre-image of E under the map

$$f: \mathbb{R}^{n \cdot n} \to S, \qquad A \mapsto {}^tAA$$

into the set S of symmetric matrices $(S = \mathbb{R}^{\frac{1}{2}n(n+1)})$.

In calculating the differential of f we consider the mapping of the paths $w(\lambda) = A + \lambda \cdot B$ through the point A with $f(A) = E$:

$$f(A + \lambda B) = E + \lambda({}^tAB + {}^tBA) + \lambda^2 \cdot {}^tBB.$$

Thus, $T_A(f)\,(\mathbb{R}^{n \cdot n})$ contains precisely all matrices of the form $({}^tAB + {}^tBA)$, where $^tAA = E$ and $B \in \mathbb{R}^{n \cdot n}$ is arbitrary. These are, however, precisely all the symmetric matrices, as one can see, if for a symmetric matrix C, one puts $B = \frac{1}{2}AC$. Therefore E is a regular point of f, $O(n) \subset \mathbb{R}^{n \cdot n}$ is a submanifold, and its codimension is dim $(S) = \frac{1}{2}n(n + 1)$. $\qquad\square$

(5.11) Definition. Let M, N be differentiable manifolds and let $L \subset N$ be a k-dimensional submanifold. A differentiable map $f: M \to N$ is called *transverse* to L if the *transversality condition*

$(Tr_p) \quad T_pf(T_pM) + T_{f(p)}L = T_{f(p)}N \quad \text{if } f(p) \in L,$

is satisfied for all $p \in M$, see Fig. 38.

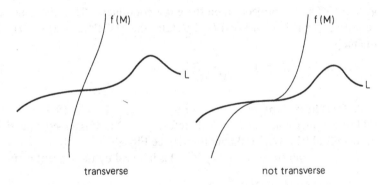

transverse   not transverse

Fig. 38

Such pictures must of course be regarded with care: the behaviour of the map cannot be read off from its image set alone.

The transversality condition imposes a requirement only on the points from the pre-image of L. For example, a map, whose image does not meet the submanifold, is certainly transverse; and if dim $M <$ codim L, then f is transverse to L if and only if $f(M) \cap L = \emptyset$, because the condition (Tr_p) cannot otherwise be satisfied. The sum of vector spaces in the transversality condition need not be *direct*, for example every map is transverse to $L = N$.

Equivalently, one can also formulate: $(Tr_p) \Leftrightarrow$ *the composition of linear maps*

$$T_pM \xrightarrow[T_pf]{} T_qN \xrightarrow[\pi]{} T_qN/T_qL, \quad \pi = projection,$$

is surjective, for $q = f(p) \in L$. The condition states that the tangent space of M is to be mapped 'as skew as possible' to that of the submanifold L.

If L is a point, then the map f is transverse to L if and only if this point is regular.

(5.12) Theorem. *If $f: M \to N$ is transverse to the k-codimensional submanifold $L \subset N$ and $f^{-1}(L) \neq \emptyset$, then $f^{-1}(L)$ is a k-codimensional submanifold of M and, for the normal bundles, one has a canonical bundle isomorphism*

$$\perp(f^{-1}L) \cong f^*(\perp L).$$

Proof. Let $f(p) = q \in L$, and in some neighbourhood V of q in suitable local coordinates let $V \cong V' \subset \mathbb{R}^n$:

$$L \cap V \cong \mathbb{R}^{n-k} \cap V',$$

where $\mathbb{R}^{n-k} \subset \mathbb{R}^n$ is given by the vanishing of the last k coordinates. Let $\pi: \mathbb{R}^n \to \mathbb{R}^k$ be the projection on these last coordinates. Then the transversality condition in a neighbourhood U of p states that $0 \in \mathbb{R}^k$ is a regular value of the map

$$U \xrightarrow[f]{} V \cong V' \xrightarrow[\pi|V']{} \mathbb{R}^k.$$

Therefore the pre-image of zero, namely, $f^{-1}(L) \cap U$, is by (5.9) a submanifold of codimension k of U and, therefore, $f^{-1}(L) \subset M$ is a k-codimensional submanifold (this is a local condition!), see Fig. 39.

The isomorphism $\perp(f^{-1}L) \to f^*(\perp L)$ is induced by the tangent map

$$Tf|f^{-1}(L): TM|f^{-1}(L) \to TN|L.$$

It induces a map $TM|f^{-1}(L) \to (TN|L)/TL$, which is linearly epimorphic on every fibre (transversality condition) and, because $T(f^{-1}L)$ obviously lies in the kernel, the map

$$\frac{TM|f^{-1}(L)}{T(f^{-1}L)} \to \frac{TN|L}{TL}$$

is an isomorphism on every fibre, which induces the required isomorphism by (3.12). $\qquad \square$

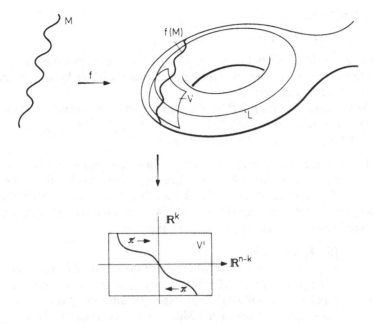

Fig. 39

The pre-image of a regular point $q \in N$ therefore has a trivial normal bundle, since it is induced from the trivial bundle $T_q(N) \to \{q\}$.

An arbitrary point of course does not need to be a regular value, nor does an arbitrary map need to be transverse; an arbitrarily prescribed closed set $A \subset M$ can arise as the pre-image of a point $q \in M$ under a differentiable mapping $M \to N$ (14.1). However, as we shall see in the following chapters, such pathological maps are 'unlikely', transversality being the usual case. The concept of transversality thus plays a decisive role in differential topology.

We close this section with a further application of the rank theorem:

(5.13) Theorem. *Let $f: M \to M$ be a differentiable map of a differentiable connected manifold into itself with $f \circ f = f$, then $f(M)$ is a closed differentiable submanifold of M.*

Proof. We have $f(M) = \{x \in M \mid f(x) = x\}$ = fixed point set of f, and this is closed.

It is sufficient to consider the map f in a neighbourhood of a point of $f(M)$. By the rank theorem it then suffices to show that the rank of f is constant in some neighbourhood of every point of $f(M)$. We first show that $rk_p f$ is constant on $f(M)$.

If $p \in f(M)$, then the differential of f at p satisfies the equation $T_p f \circ T_p f = T_p f$; therefore as above

$$\text{image } (T_p f) = \{v \in T_p M \mid T_p f(v) = v\} = \text{kernel } (\text{Id} - T_p f),$$

and thus, in particular,

$$rk_p f + rk(\mathrm{Id} - T_p f) = \dim M$$

for all $p \in f(M)$. Since both ranks on the left side can only increase in a neighbourhood of a point, $rk_p f$ is locally constant on $f(M)$, hence constant because $f(M)$ is connected.

Now let $rk_p f = r$ for $p \in f(M)$, then there is an open neighbourhood U of $f(M)$, so that $rk_q f \geqslant r$ for all $q \in U$. But $rk_q f = rk_q(f \circ f) = rk(T_{f(q)} f \cdot T_q f) \leqslant rk_{f(q)} f = r$, therefore $rk_q f$ is constant on U.   □

In general, if $A \subset X$ and $f: X \to A$ is a mapping, so that $f|A = \mathrm{Id}_A$, that is, a mapping which throws X onto A, keeping each point of A fixed, then one calls f a *retraction*. We have thus shown that the image of a differentiable retraction is a differentiable submanifold. A continuous (non-differentiable) retraction can, however, have very wild image sets.

(5.14) Exercises

1 Let $\mathbb{R} + \mathbb{R}$ be a differentiable sum of the manifold $\mathbb{R}$ with itself (1.8), and let $f: \mathbb{R} + \mathbb{R} \to \mathbb{R}^2$ be the map with the components $f_1(x) = (x, 0)$ and $f_2(y) = (0, \exp(y))$. Show that f is an injective immersion, but not an embedding, and draw a sketch of the image.

2 Let the map $f: \mathbb{R} + S^1 \to \mathbb{C}$ have the components

$$f_1(t) = (1 + \exp(t)) \cdot \exp(it),$$

$$f_2(\exp(it)) = \exp(it), \quad \text{with} \quad S^1 = \{z \in \mathbb{C} \mid |z| = 1\}.$$

Show that f is an injective immersion, but not an embedding and draw a sketch.

3 (a) Show that if $c \in \mathbb{R}$ is irrational, then the subgroup generated by $\exp(2\pi i c)$ is dense in $S^1 = \{z \in \mathbb{C} \mid |z| = 1\}$.
   (b) The map $\mathbb{R} \to \mathbb{C} \times \mathbb{C}$

   $$t \mapsto (\exp(ait), \exp(bit))$$

   is an immersion if $b \neq 0$; show that if a/b is irrational, then it is injective and the image is dense in $S^1 \times S^1 \subset \mathbb{C}^2$.

4 Let A be a symmetric real $(n \times n)$-matrix, and $0 \neq b \in \mathbb{R}$, show that the *quadric*

   $$M = \{x \in \mathbb{R}^n \mid {}^t x A x = b\}$$

   is an $(n-1)$-dimensional submanifold of $\mathbb{R}^n$.

5 For an integer $d \geqslant 0$ the *Brieskorn manifold* $W^{2n-1}(d)$ is defined as the set of points $(z_0, \ldots, z_n) \in \mathbb{C}^{n+1}$, which satisfy the equations

   $$z_0^d + z_1^2 + \ldots + z_n^2 = 0$$

   $$z_0 \bar{z}_0 + z_1 \bar{z}_1 + \ldots + z_n \bar{z}_n = 2$$

   Show that $W^{2n-1}(d)$ is a $(2n-1)$-dimensional manifold.

6 Let $\mathbb{CP}^n$ be a complex projective space, and

$$H(m,n) = \left\{(z,w) \in \mathbb{CP}^m \times \mathbb{CP}^n \middle| \sum_{i=0}^m z_i w_i = 0\right\}$$

for $m \leqslant n$, where $z = [z_0, \ldots, z_m]$ and $w = [w_0, \ldots, w_n]$ are homogeneous coordinates. Show that $H(m,n)$ is a $2(m+n-1)$-dimensional manifold. Corresponding manifolds are also obtained from the real projective spaces. They are called *Milnor manifolds*.

7 Show that the manifold of orthogonal matrices $O(n)$ is compact, the group operations

$$O(n) \times O(n) \to O(n) \quad \text{(multiplication)},$$

$$O(n) \to O(n), \quad A \mapsto A^{-1},$$

differentiable, and that $O(n)$ has two connected components.

8 A k-frame in $\mathbb{R}^n$ is an orthonormal k-tuple $(v_1, \ldots, v_k)$ of vectors in $\mathbb{R}^n$. The set $V_n^k \subset \mathbb{R}^n \times \ldots \times \mathbb{R}^n$ (k factors) of k-frames in $\mathbb{R}^n$ is called a *Stiefel manifold*. Show that V_n^k is a compact differentiable manifold of dimension $n \cdot k - \frac{1}{2} \cdot k \cdot (k+1)$.

9 Show that the set $U(n)$ of unitary matrices, considered as a subset of $O(2n)$, is a submanifold of $O(2n)$ of dimension n^2.

10 Let $f: M \to N$ be a differentiable retraction and $p \in f(M)$. Show that there is a local coordinate system around p, in which f is given by

$$(x_1, \ldots, x_r, \ldots, x_n) \mapsto (x_1, \ldots, x_r, 0, \ldots, 0).$$

Note that here, unlike in the rank theorem, one cannot choose charts independently in the image and pre-image manifolds!

11 Let M, N, L be differentiable manifolds, and

$$M \xrightarrow{f} N \xleftarrow{g} L$$

differentiable maps, so that for every point $p \in M$ and $q \in L$ with $f(p) = g(q) = r \in N$, we have

$$T_p f(T_p M) + T_q g(T_q L) = T_r(N).$$

Show that the fibre product of f and g:

$$\{(p,q) \in M \times L \,|\, f(p) = g(q)\}$$

is a differentiable manifold.

6
Sard's theorem

The aim of this chapter is the proof of the following theorem.

(6.1) **Sard's theorem.** *The set of critical values of a differentiable mapping of manifolds has Lebesgue measure zero.*

In particular, if $f: M \to \mathbb{R}^n$ is differentiable, then for almost all b the set $f^{-1}\{b\} \subset M$ is an n-dimensional submanifold; in other words the differentiable system of equations on M

$$
\begin{aligned}
f_1(x) &= b_1 \\
\vdots \quad & \quad \vdots \\
f_n(x) &= b_n
\end{aligned}
$$

has (given f) for nearly every choice of b_i an n-codimensional submanifold of M as its solution set (5.9).

We now come to more detailed explanations:

(6.1) **Definition.** A subset $C \subset \mathbb{R}^n$ has *measure zero (almost every point is not in C)*, if for every $\epsilon > 0$ there is a sequence of cubes $W_i \subset \mathbb{R}^n$ with $C \subset \bigcup_{i=1}^{\infty} W_i$ and $\sum_{i=1}^{\infty} |W_i| < \epsilon$. Here, $|W|$ is the volume of the cube W, that is $|W| = (2a)^n$ if $W = \{x \mid |x_i - x_i^0| \leqslant a\}$.

A countable union of sets of measure zero again has measure zero, for if we have $C \subset \bigcup_{\nu=1}^{\infty} C_\nu$ and $C_\nu \subset \bigcup_{i=1}^{\infty} W_i^\nu$ with $\sum_{i=1}^{\infty} |W_i^\nu| < \epsilon/2^\nu$, then $C \subset \bigcup_{i,\nu} W_i^\nu$ and $\sum_{i,\nu} |W_i^\nu| < \epsilon$. For similar reasons it does not matter if one takes open or closed cubes, rectangular blocks, or balls.

(6.2) **Lemma.** *Let $U \subset \mathbb{R}^m$ be open, $C \subset U$ a set of measure zero, and let $f: U \to \mathbb{R}^m$ be differentiable, then $f(C)$ also has measure zero.*

Proof. Since U is the union of a sequence of compact balls, one may assume that C is contained in a compact ball, and that the cubes of a covering of C according to (6.1) are also contained in a somewhat larger ball $K \subset U$. The mean value theorem of differential calculus provides an estimate

56

$$f(x + h) = f(x) + R(x, h)$$

$$|R(x, h)| \leqslant c |h|$$

for $x, x + h \in K$, for some constant c. If, therefore, a cube $W \subset K$ has edge length a, $|x - x^0| \leqslant \sqrt{m} \cdot a$ for $x \in W$, and $|f(x) - f(x^0)| \leqslant c \cdot \sqrt{m} \cdot a$. Thus $f(W)$ lies in a cube of volume $(2 \cdot \sqrt{m} \cdot c)^m |W|$ and, because the constant $(2 \cdot \sqrt{m} \cdot c)^m$ is independent of the cube, the assertion follows. $\qquad\square$

This lemma makes it meaningful also to speak of sets of measure zero in a differentiable manifold.

(6.3) Definition. A subset C of a differentiable manifold M has *measure zero* if for every chart $h: U \to U' \subset \mathbb{R}^m$ the set $h(C \cap U) \subset \mathbb{R}^m$ has measure zero.

Since a manifold has a countable base for its topology, from every atlas one can choose a subatlas with countably many charts (Kelley [8], chapter 1, theorem 15, p. 49); if one applies lemma (6.2) to the chart transformations in such an atlas, then it follows that C has measure zero if for all charts h_α of a chosen fixed atlas $h_\alpha(C \cap U_\alpha)$ has measure zero in $\mathbb{R}^m$.

A corresponding definition for a topological manifold has no meaning because non-differentiable homeomorphisms can map a measure mero set onto a set of positive measure (an example of this cannot be simply given).

After the introduction of charts it is only necessary to carry out the proof that a set has measure zero for subsets of $\mathbb{R}^n$. Here the following special case of Fubini's theorem provides an induction procedure:

(6.4) Theorem (Fubini). *Let* $\mathbb{R}_t^{n-1} := \{x \in \mathbb{R}^n | x_n = t\}$; *let* $C \subset \mathbb{R}^n$ *be compact and* $C_t := C \cap \mathbb{R}_t^{n-1}$ *have measure zero in* $\mathbb{R}_t^{n-1} \cong \mathbb{R}^{n-1}$ *for all* $t \in \mathbb{R}$. *Then* C *has measure zero in* $\mathbb{R}^n$.

Proof (following Sternberg [9]). We use the following elementary

Proposition. *An open covering of the interval* $[0, 1]$ *by subintervals contains a finite covering* $[0, 1] = \cup_{j=1}^k I_j$ *with* $\Sigma_{j=1}^k |I_j| \leqslant 2$.

Proof of the above. One chooses a finite subcovering, from which it is not possible to exclude any further interval. Then every point of $[0, 1]$ lies only in, at most, two intervals of this covering: if it were to lie in three, then one of these would have the smallest initial point and one would have the largest end point, a further one would be superfluous. $\qquad\square$

Now we come to Fubini's theorem. W.l.o.g., let $C \subset \mathbb{R}^{n-1} \times [0, 1]$, and C_t have measure zero in $\mathbb{R}^{n-1} \times t$ for all $t \in [0, 1]$. For every $\epsilon > 0$ we find a covering of C_t by open cubes W_t^i in $\mathbb{R}_t^{n-1}$ with volume sum $< \epsilon$. Let W_t be the projection of $\cup_i W_t^i \subset \mathbb{R}_t^{n-1}$ on the first factor $\mathbb{R}^{n-1}$ of $\mathbb{R}^{n-1} \times [0, 1]$, see Fig. 40. If x_n is the last coordinate, then, for fixed t, the function $|x_n - t|$ is continuous on C, it vanishes precisely on C_t and outside of $W_t \times [0, 1]$ it

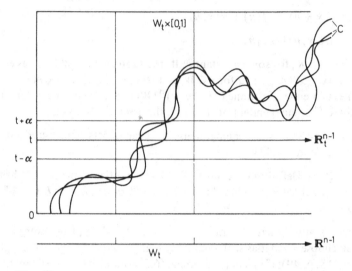

Fig. 40

attains a minimum value α because C is compact. Thus we have

$$\{x \in C \,|\, x_n - t| < \alpha\} \subset W_t \times I_t \text{ with } I_t = (t - \alpha, t + \alpha).$$

The various intervals I_t constructed like this cover $[0, 1]$ and, according to the proposition, from these we may choose a finite subcovering $\{I_j \,|\, j = 1, \ldots, k\}$ of volume sum $\leqslant 2$. Here $I_j = I_{t_j}$ for some $t_j \in [0, 1]$. The rectangular blocks

$$\{W^i_{t_j} \times I_j \,|\, j = 1, \ldots, k; i \in \mathbb{N}\}$$

cover C and have volume sum $< 2\epsilon$.   ☐

(6.5) Remark. The hypothesis that C is compact may be easily weakened; it is clearly sufficient that C be a countable union of compact sets. In particular, this holds for closed sets, open sets, images of sets of this class under continuous maps, countable unions and finite intersections of such sets. This class will satisfy us.

With this we come to the proof of Sard's theorem (see Milnor [6]). Following the introduction of charts, by definition (6.3) one has to show the following:

Let $U \subset \mathbb{R}^n$ be open, $f: U \to \mathbb{R}^p$ differentiable and let $D \subset U$ be the set of critical points of f, then $f(D) \subset \mathbb{R}^p$ has measure zero.

Proof. This is by induction on n; for $n = 0$, $\mathbb{R}^n$ is a point, $f(U)$ is, at most, one point and the theorem holds.

For the induction step let $D_i \subset U$ be the set of points $x \in U$, at which all partial derivatives of order $\leqslant i$ vanish. The D_i clearly form a decreasing sequence of closed sets

$$D \supset D_1 \supset D_2 \supset \dots,$$

and we show

(a) $f(D - D_1)$ has measure zero,
(b) $f(D_i - D_{i+1})$ has measure zero and,
(c) $f(D_k)$ has measure zero for sufficiently large k.

Note that all the sets appearing here fall into the class to which by (6.5) we may apply Fubini's theorem; further, it suffices in each case to show that each point $x \in D - D_1$ (... respectively) possesses a neighbourhood V so that $f(V \cap (D - D_1))$ has measure zero, for $D -- D_1$ (... respectively) is covered by countably many such neighbourhoods.

Proof (a). One can assume $p \geqslant 2$ since for $p = 1, D = D_1$. Let $x \in D - D_1$; as $x \notin D_1$, some partial derivative of f does not vanish at the point x, we may therefore assume that $\partial f / \partial x_1(x) \neq 0$; then by (5.5) the map

$$h: U \to \mathbb{R}^n, (x_1, \dots, x_n) \mapsto (f_1(x), x_2, \dots, x_n)$$

is not singular at the point x; so its restriction to a neighbourhood V of x is a chart $h: V \to V'$, and the transformed map $g := f \circ h^{-1}$ has the form

$$g: (z_1, \dots, z_n) \mapsto (z_1, g_2(z), \dots, g_p(z))$$

locally about $h(x)$. The mapping takes the hyperplane $\{z \mid z_1 = t\}$ into the plane $\{y \mid y_1 = t\}$; let

$$g^t: (t \times \mathbb{R}^{n-1}) \cap V' \to t \times \mathbb{R}^{p-1}$$

be the restriction of g. Then a point from $(t \times \mathbb{R}^{n-1}) \cap V'$ is critical for g if and only if it is critical for g^t, since g has the Jacobi matrix

$$Dg = \begin{array}{|c|c|} \hline 1 & 0 \\ \hline ? & Dg' \\ \hline \end{array}$$

However, by the inductive hypothesis, the set of critical values of g^t has measure zero in $t \times \mathbb{R}^{n-1}$, thus the set of critical values of g has an intersection of measure zero with each hyperplane $\{y \mid y_1 = t\}$. Hence, by Fubini's theorem, it itself also has measure zero, and (a) is proved.

Proof (b). We proceed similarly. For each point

$x \in D_k - D_{k+1}$ there exists some $(k + 1)$st derivative which does not vanish at the point x. We may suppose that

$$\partial^{k+1} f_1 / \partial x_1 \partial x_{v_1}, \ldots, \partial x_{v_k}(x) \neq 0.$$

Let $w: U \to \mathbb{R}$ be the function

$$w = \partial^k f_1 / \partial x_{v_1}, \ldots, \partial x_{v_k},$$

then therefore $w(x) = 0$, $\partial w / \partial x_1(x) \neq 0$, and as before the map

$$h: x \mapsto (w(x), x_2, \ldots, x_n),$$

defines a chart $h: V \to V'$ about x, and

$$h(D_k \cap V) \subset 0 \times \mathbb{R}^{n-1} \subset \mathbb{R}^n.$$

We may therefore again consider the transformed map $g := f \circ h^{-1}: V' \to \mathbb{R}^p$, and its restriction $g^0: (0 \times \mathbb{R}^{n-1}) \cap V' \to \mathbb{R}^p$, which by the inductive hypothesis has a set of critical values of measure zero. However, each point from $h(D_k \cap V)$ is critical for g^0 because all partial derivatives of g, hence also of g^0, of order $\leqslant k$, in particular of first order, vanish. Therefore $f(D_k \cap V) = g \circ h(D_k \cap V)$ has measure zero.

Proof (c). Let $W \subset U$ be a cube with edges of length a, and let $k > (n/p) - 1$, then we shall show that $f(W \cap D_k)$ has measure zero. Since U is a countable union of cubes, this will be sufficient. The Taylor formula yields an estimate

$$f(x + h) = f(x) + R(x, h), \quad |R(x, h)| \leqslant c \cdot |h|^{k+1},$$

for $x \in D_k \cap W$ and $x + h \in W$, where the constant c is fixed for given f and W.

Now decompose W into r^n cubes with edges of length a/r. If W_1 is a cube of the decomposition, which contains a point $x \in D_k$, then each point from W_1 can be written as $x + h$ with

$$|h| \leqslant \frac{\sqrt{n} \cdot a}{r}.$$

From the remainder estimate above, $f(W_1)$ lies in a cube with edges of length

$$2 \cdot c \cdot \frac{(\sqrt{n} \cdot a)^{k+1}}{r^{k+1}} = \frac{b}{r^{k+1}},$$

with a constant b, depending only on W and f and not on the decomposition. All these cubes together have a combined volume $s \leqslant r^n \cdot b^p / r^{p(k+1)}$ and, for $p(k + 1) > n$, this expression converges to zero as r increases. Hence, by choice of a sufficiently fine decomposition, the combined volume can be made arbitrarily small.   □

The most important consequence of Sard's theorem is the older result of Brown, which we wish to state separately:

(6.6) Consequence. *The regular values of a differentiable map* $f: M \to N$ *are dense in* N.

(6.7) Exercises

1 Let $f: M \to N \times \mathbb{R}^n$ be a differentiable map; show that for each $\epsilon > 0$ there exists a vector $v \in \mathbb{R}^n$ with $|v| < \epsilon$, so that the map

$g: M \to N \times \mathbb{R}^n, \quad x \mapsto f(x) + v$

is transverse to the submanifold $N \times 0 \subset N \times \mathbb{R}^n$.

2 Show that, if $M^n \subset \mathbb{R}^p$ is a differentiable submanifold, then there exists a hyperplane in $\mathbb{R}^p$, which cuts M^n transversally.

3 Show that there is no surjective differentiable map $\mathbb{R}^n \to \mathbb{R}^{n+1}$.

4 Let M^n be a compact manifold, $f: M^n \to \mathbb{R}^{n+1}$ differentiable, and $0 \notin f(M)$. Show that there exists a line through the origin of $\mathbb{R}^{n+1}$, which only meets finitely many points of $f(M^n)$.

5 Let $f: M \to \mathbb{R}^p$ be a differentiable map and $N \subset \mathbb{R}^p$ a differentiable submanifold. Show that for each $\epsilon > 0$ there exists $v \in \mathbb{R}^p$, with $|v| < \epsilon$, so that the map $M \to \mathbb{R}^p, x \mapsto f(x) + v$ is transverse to N.
   Hint: consider the map $M \times N \to \mathbb{R}^p, (x, y) \mapsto y - f(x)$.

6 For a differentiable map $f: M \to N$ let

$$\Sigma^i(f) := \{p \in M \mid rk_p f = i\}.$$

Let $f: \mathbb{R}^m \to \mathbb{R}^n$ be differentiable and $\epsilon > 0$. Show that there exists a linear map $\alpha: \mathbb{R}^m \to \mathbb{R}^n$ of norm $< \epsilon$, so that $\Sigma^i(f + \alpha)$ is a differentiable submanifold of $\mathbb{R}^m$.
   Hint: apply exercise 5 to Df and use (1.11, 16).

7 Let $f: \mathbb{R}^m \to \mathbb{R}^n$ be differentiable and $m \leqslant 2n$. Show that for each $\epsilon > 0$ there exists a linear map $\alpha: \mathbb{R}^m \to \mathbb{R}^n$ of norm $< \epsilon$, so that the map $f + \alpha: \mathbb{R}^m \to \mathbb{R}^n$ is an immersion.
   Hint: this is a side result of the solution to exercise 6.

8 Let $M^k \subset \mathbb{R}^{n+1}$ be a compact submanifold and $n \geqslant 2k$. Show that, for the projection $\pi: \mathbb{R}^{n+1} \to H^n$ onto a suitable hyperplane H of $\mathbb{R}^{n+1}$, the restriction $\pi|M: M \to H$ is an immersion.
   Hint: consider the $(2k - 1)$-dimensional manifold PTM, whose elements are the 1-dimensional subspaces of the tangent spaces of M, and study the canonical map $PTM \to \mathbb{R}P^n$.

9 Let $M^k \subset \mathbb{R}^{n+1}$ be a compact submanifold and $n \geqslant 2k + 1$. Show that, for the projection $\pi: \mathbb{R}^{n+1} \to H^n$ onto a suitable hyperplane H of $\mathbb{R}^{n+1}$, the restriction $\pi|M: M \to H$ is an embedding.

7
Embedding

What we have studied up to now — apart from the tangent bundle — is essentially the local structure of differentiable manifolds, and at first it is not obvious that between two manifolds there can ever exist non-trivial maps, and that everything which one intuitively describes as 'smooth' can also be realised by means of differentiable maps. The essential technical tool for the passage from local to global is the partition of unity which we now manufacture.

(7.1) Lemma. *Let M be a differentiable manifold and $\mathfrak{U} = \{U_\lambda \,|\, \lambda \in \Lambda\}$ an open covering of M. Then there exists an atlas $\mathfrak{A} = \{h_\nu \colon V_\nu \to V'_\nu \,|\, \nu \in \mathbb{N}\}$ of M with the following properties:*

(a) *$\{V_\nu \,|\, \nu \in \mathbb{N}\}$ is a locally finite refinement of $\{U_\lambda \,|\, \lambda \in \Lambda\}$,*
(b) *$V'_\nu = \{x \in \mathbb{R}^m \,|\, |x| < 3\} =: K(3)$,*
(c) *The sets $W_\nu := h_\nu^{-1}\{x \in \mathbb{R}^m \,|\, |x| < 1\} = h_\nu^{-1} K(1)$ still cover M.*

Such an atlas is called a good atlas subordinate to the covering $\mathfrak{U}$.

Proof. Because M is locally compact with a countable basis, we can easily find a sequence of compact subsets A_i, so that $A_i \subset \mathring{A}_{i+1}$ and $\cup_{i=1}^\infty A_i = M$. (Choose a countable cover $\{C_n\}_{n \in \mathbb{N}}$ of M by compact sets, and then choose $A_1 = C_1$ and A_n as a compact neighbourhood of $A_{n-1} \cup C_n$.) Now for each i we may choose finitely many charts $h_\nu \colon V_\nu \to K(3)$, so that $V_\nu \subset \mathring{A}_{i+2} - A_{i-1}$ and $V_\nu \subset U_\lambda$ for some λ, and so that the sets $W_\nu = h_\nu^{-1}(K(1))$ still form a cover of $A_{i+1} - \mathring{A}_i$. This follows easily, since this set is compact and has $\mathring{A}_{i+2} - A_{i-1}$ as an open neighbourhood (see Fig. 41). All these charts for all $i \in \mathbb{N}$ together form the sought for atlas

$\square$

Next we recall that the function (illustrated in Fig. 42)

$$\lambda \colon \mathbb{R} \to \mathbb{R}, \quad t \mapsto \begin{cases} 0 & \text{for } t \leq 0 \\ \exp(-t^{-2}) & \text{for } t > 0 \end{cases}$$

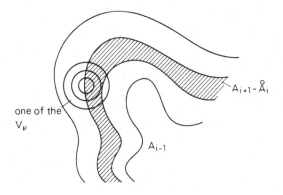

Fig. 41

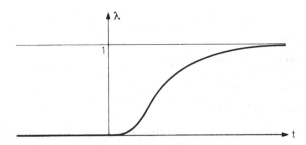

Fig. 42

is infinitely differentiable, that $0 \leq \lambda \leq 1$, and that $\lambda(t) = 0 \Leftrightarrow t \leq 0$. For $t > 0$ the derivatives of λ have the form $q(t) \cdot \exp(-t^{-2})$, where q is a rational function, and they therefore converge to zero as t goes to zero. Now let $\epsilon > 0$ and $\phi_\epsilon(t) = \lambda(t) \cdot (\lambda(t) + \lambda(\epsilon - t))^{-1}$ (Fig. 43), then ϕ_ϵ is differentiable, $0 \leq \phi_\epsilon \leq 1$, and $\phi_\epsilon(t) = 0 \Leftrightarrow t \leq 0$, and $\phi_\epsilon(t) = 1 \Leftrightarrow t \geq \epsilon$. For the ball

$$K(r) = \{x \in \mathbb{R}^n \mid |x| < r\}, \quad r > 0,$$

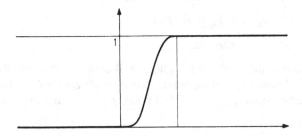

Fig. 43

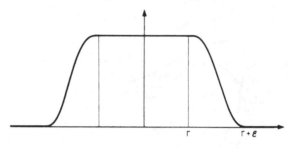

Fig. 44

we can therefore find the differentiable *bump function* (Fig. 44)

(7.2) $\psi: \mathbb{R}^n \to \mathbb{R}$

$$\psi(x) = 1 - \phi_\epsilon(|x| - r)$$

with the properties:

$$0 \leqslant \psi(x) \leqslant 1 \quad \text{for all} \quad x \in \mathbb{R}^n,$$

$$\psi(x) = 1 \Leftrightarrow x \in \overline{K(r)},$$

$$\psi(x) = 0 \Leftrightarrow |x| \geqslant r + \epsilon.$$

About the point $x = 0$, where $|x|$ is not differentiable, ψ is locally constant and therefore differentiable.

If one composes such a bump function with a suitable chart, then one obtains a function $\psi \cdot h: U \to \mathbb{R}$ on the chart domain of a manifold, and because this function vanishes outside $h^{-1}K(r + \epsilon) \subset U$, one can extend it (by 0 on $M - U$) as a differentiable function over the whole manifold M.

(7.3) Theorem. *For every open covering of a differentiable manifold, there exists a subordinate differentiable partition of unity.*

Proof. Using (7.1) we may choose a good atlas $\mathfrak{A}$ subordinate to the cover $\mathfrak{U}$ of M, also a bump function ψ for the ball $K(1)$ with $\psi|K(1) = 1$, $\psi(x) = 0$ for $|x| \geqslant 2$. Define the function ψ_ν on M by

$$\psi_\nu = \begin{cases} \psi \cdot h_\nu & \text{on } V_\nu = h_\nu^{-1}K(3) \\ 0 & \text{otherwise} \end{cases}$$

Then ψ_ν is differentiable, and $s = \sum_{\nu=1}^{\infty} \psi_\nu$ is well defined and differentiable, since the family $\{\text{Supp} (\psi_\nu)\}$ of supports is locally finite and differentiability is a local property. Besides, $s(p) \neq 0$ for all points $p \in M$, so that the functions

$$\phi_\nu := (1/s)\psi_\nu$$

form the sought for partition of unity.   □

An easy consequence:

(7.4) Remark. If A_0, A_1 are disjoint closed subsets of the differentiable manifold M, then there exists a differentiable function *(separating function)* $\phi: M \to \mathbb{R}, 0 \leqslant \phi \leqslant 1$, so that $\phi|A_0 = 0, \phi|A_1 = 1$.

Proof. Let $\{\phi_\nu | \nu \in \mathbb{N}\}$ be a partition of unity subordinate to the covering by the sets $U_i = M - A_i$ and put

$$\phi = \sum_{\nu \in K} \phi_\nu$$

with $\nu \in K$ if and only if Supp $(\phi_\nu) \subset U_1$. $\square$

In what follows, we shall concern ourselves with approximations to given maps with 'nice' properties (embeddings, transverse maps, etc.). In doing this we must ensure that for the approximation, not only the values of the function, but also the values of the partial derivatives, undergo no more than a small change. However, we do not wish to involve ourselves unnecessarily with the appropriate topologies on the set of differentiable maps $C^\infty(M, N)$, and restrict ourselves to the bare minimum.

(7.5) Definition. Let $U \subset \mathbb{R}^m$ be open and $K \subset U$ compact; let $f \in C^\infty(U)$, then set

$$|f|_K := \max \{|f(x)| \,|\, x \in K\} + \sum_{\nu=1}^{m} \max \{|\partial f/\partial x_\nu(x)| \,|\, x \in K\}.$$

If $f = (f_1, \ldots, f_n): U \to \mathbb{R}^n$, then $|f|_K := \max \{|f_\nu|_K\}$.

It is straightforward to check that $|f|_K$ defines a seminorm on $C^\infty(U)$, that is,

$$|f + g|_K \leqslant |f|_K + |g|_K,$$
$$|\lambda f|_K = \lambda |f|_K \quad \text{for} \quad \lambda > 0,$$
$$|f \cdot g|_K \leqslant |f|_K \cdot |g|_K.$$

Furthermore, for $K \subset L, |f|_K \leqslant |f|_L$, but clearly it is possible that $|f|_K = 0$ without $f = 0$ (but $f|K = 0$).

In particular this seminorm makes $C^\infty(U, \mathbb{R}^n)$ into a *topological space* $C^\infty(U, \mathbb{R}^n)_K$; ϵ-neighbourhoods with respect to the seminorm $|f|_K$ form a neighbourhood basis.

(7.6) Lemma. Let U be open in $\mathbb{R}^m$ and $K \subset U$ compact; the set of differentiable maps $f: U \to \mathbb{R}^n$, which have rank m at all points of K, is open in $C^\infty(U, \mathbb{R}^n)_K$, and is dense in the case of $2m \leqslant n$.

Proof. The condition $rk_x f = m$ means that the Jacobi matrix Df_x has rank m, or that the map $K \to \mathbb{R}^{m \cdot n}, x \mapsto Df_x$ has image contained in the open set of matrices of rank $\geqslant m$. If now $|f-g|_K$ is sufficiently small, then it follows that $|Df_x - Dg_x|$ is so small on K that $Dg_x|K$ also maps into this open set (see (7.5)).

Now let $2m \leqslant n$, $\epsilon > 0$, and let the vectors $\partial f/\partial x_i$ for $i = 1, \ldots, s < m$ be already linearly independent at each point of U, then we find a map g with $|f-g|_K < \epsilon$ so that the vectors $\partial g/\partial x_i, i = 1, \ldots, s+1$ are linearly independent at each point. The result will then follow by induction. To this end we consider the map

$$\phi: \mathbb{R}^s \times U \to \mathbb{R}^n, (\lambda_1, \ldots, \lambda_s, x) \mapsto \sum_{j=1}^{s} \lambda_j \frac{\partial f}{\partial x_j}(x) - \frac{\partial f}{\partial x_{s+1}}(x).$$

For $s < m$ dim $(\mathbb{R}^s \times U) = s + m < 2m \leqslant n$, so by Sard's theorem we can find a point $a = (a_1, \ldots, a_n) \in \mathbb{R}^n$ of arbitrarily small norm with $a \notin \phi(\mathbb{R}^s \times U)$. Now, set

$$g(x) = f(x) + x_{s+1} \cdot a,$$

then $\partial g/\partial x_i = \partial f/\partial x_i$ for $i \leqslant s$, and $\partial g/\partial x_{s+1} = \partial f/\partial x_{s+1} + a$. A linear relation

$$\sum_{j=1}^{s} \lambda_j \frac{\partial g}{\partial x_j} = \frac{\partial g}{\partial x_{s+1}}$$

is satisfied nowhere in U, for this would imply that

$$\sum_{j=1}^{s} \lambda_j \frac{\partial f}{\partial x_j} - \frac{\partial f}{\partial x_{s+1}} = a. \qquad \square$$

In this proof only the trivial case (6.2) of Sard's theorem is used. Another proof depends on (6.7, 7). From this local result by means of a good atlas we can cobble the appropriate global result together.

(7.7) Immersion theorem (H. Whitney). *Let M^m be a differentiable manifold, $\delta: M \to \mathbb{R}$ an everywhere strictly positive continuous function and $f: M \to \mathbb{R}^n$ a differentiable map with $2m \leqslant n$. Let $A \subset M$ be closed and $rk_p f = m$ for all $p \in A$. Then there exists an immersion $g: M \to \mathbb{R}^n$, with $g|A = f|A$ and $|g(p) - f(p)| < \delta(p)$ for all $p \in M$.*

In other words, one cannot only find an immersion $M \to \mathbb{R}^n$, but one can also always approximate a given map by an immersion where the 'nearness' δ of the approximation can be prescribed by an arbitrary continuous positive function.

One can express such statements about approximation more elegantly in terms of a topology on $C^\infty(M, N)$.

(7.8) Definition. *Let U be open in $M \times N$ and V_U the set of $g \in C^\infty(M, N)$ for which the graph $\{(p, g(p))| p \in M\}$ lies completely in U.*

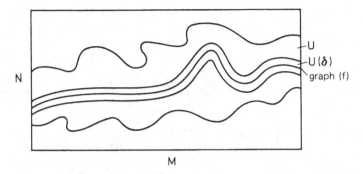

Fig. 45

The C^0-topology (which is the only one considered in this book) on $C^\infty(M, N)$ has the sets V_U as a basis of open sets, see Fig. 45.

If one chooses a metric d on N, and a differentiable manifold admits a metric (7.12), then, given a neighbourhood V_U of $f \in C^\infty(M, N)$, one easily constructs a continuous function $\delta: M \to \mathbb{R}$, $\delta > 0$, so that

$$U(\delta) := \{(p, q) \mid d(f(p), q) < \delta(p)\} \subset U.$$

(Let $\{\phi_n \mid n \in \mathbb{N}\}$ be a partition of unity on M with compact supports, and $\delta_n > 0$ be such that $(p, q) \in U$ for $p \in \mathrm{Supp}\,(\phi_n)$, and $d(f(p), q) < \delta_n$; set $\delta = \sum_{n=1}^{\infty} \delta_n \phi_n$.) Therefore, one can certainly restrict oneself, as in the theorem, to consideration of special neighbourhoods $V_\delta := V_{U(\delta)}$ of a map f. Moreover, the C^0-topology does not depend on the choice of metric and, if M is compact, one can choose δ to be constant (topology of uniform convergence). With the help of locally finite atlases on M and N one can also introduce topologies on $C^\infty(M, N)$ which describe the convergence of the higher derivatives in the same way that the C^0-topology describes the convergence of the function values. But we do not want to go into this (see Narasimham [7]).

The immersion theorem thus says that immersions are dense in $C^\infty(M, \mathbb{R}^n)$, if $2m \leqslant n$; also one does not need to disturb the map f on any closed set, where it already has maximal rank.

Proof of the immersion theorem. Because locally the rank of f cannot decrease (5.3), there exists an open neighbourhood U of A, so that $rk_p(f) = m$ for all $p \in U$. For the cover $\{(M - A), U\}$ of M we choose, using (7.1), a subordinate good atlas $\{h_\nu: V_\nu \to K(3) \mid \nu \in \mathbb{Z}\}$; the sets $W_\nu = h_\nu^{-1} K(1)$ still cover M. We set $U_\nu = h_\nu^{-1} K(2)$ and so arrange the numbering that $V_\nu \subset U$ if and only if $\nu < 1$. Only in the chart domains V_ν with positive index will g be different from f. Inductively, we construct maps $g_\nu: M \to \mathbb{R}^n$, $\nu \geqslant 0$, with the following properties:

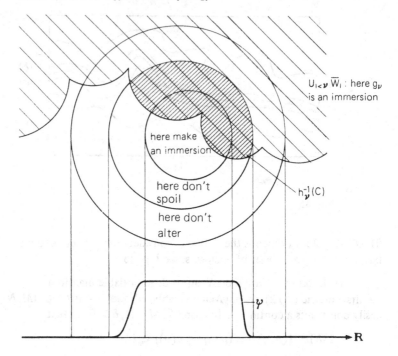

$U_{l<\nu} \bar{W}_l$: here g_ν is an immersion

here make an immersion

here don't spoil

here don't alter

$h_\nu^{-1}(C)$

ψ

R

Fig. 46

(a) $g_0 = f$,
(b) $g_\nu(x) = g_{\nu-1}(x)$ for $x \notin U_\nu$,
(c) if $d = \min \{\delta(x)|x \in \bar{U}_\nu\}$, then $|g_\nu(x) - g_{\nu-1}(x)| < \epsilon_\nu :=$
    $d/2^\nu$ for all $x \in M$,
(d) g_ν has rank m on $\bigcup_{i \leqslant \nu} \bar{W}_i$.

Having done this, we set $g = \lim_{\nu \to \infty} g_\nu$. Because the covering $\{U_\nu\}$ is locally finite (b) implies that $g_{\nu+1}(x) = g_\nu(x)$ for almost all ν, and the sequence g_ν converges to a differentiable map g, which, by (a) and the numbering of our atlas, coincides with f on A. Locally for large ν, g agrees with g_ν and, by (d), therefore has maximal rank m. Finally by (c)

$$|g - f| = |g - g_0| \leqslant \delta \sum_\nu 2^{-\nu} = \delta.$$

We now come to the construction of the sequence g_ν, illustrated in Fig. 46. For this, by (7.2), we choose a bump function $\psi \colon \mathbb{R}^m \to \mathbb{R}$ for $K(1)$ with support in $K(2)$, and a bound s, so that $|\psi|_K \leqslant s$ for $K = \overline{K(2)}$, hence for all K. Now consider the map

$$g_{\nu-1} \cdot h_\nu^{-1} \colon K(3) \to \mathbb{R}^n.$$

It has rank m on the compact set $C := h_\nu(\bar{U}_\nu \cap \cup_{i<\nu} \bar{W}_i) \subset \overline{K(2)}$ and, by the local result (7.6), the same holds for every map $q: K(3) \to \mathbb{R}^n$ with $|g_{\nu-1} \circ h_\nu^{-1} - q|_C < \eta$ for some suitable $\eta > 0$. Using (7.6) again, we find such a q, which has rank m on $\overline{K(2)}$, and such that $|g_{\nu-1} \circ h_\nu^{-1} - q|_K < \zeta < \min \{\eta \cdot s^{-1}, \epsilon_\nu\}$.

We set

$$g_\nu(x) = \begin{cases} g_{\nu-1}(x) + \psi \circ h_\nu(x) \cdot (q \circ h_\nu(x) - g_{\nu-1}(x)) & \text{for } x \in V_\nu \\ g_{\nu-1}(x) & x \notin \bar{U}_\nu, \text{see Fig. 47.} \end{cases}$$

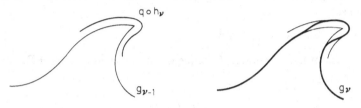

Fig. 47

The definitions agree on the open intersection of the domains of definition since there $\psi \circ h_\nu = 0$. Moreover,

$$|g_\nu h_\nu^{-1} - g_{\nu-1} \circ h_\nu^{-1}|_C \leqslant s \cdot \zeta < \eta$$

and therefore $g_\nu \circ h_\nu^{-1}$ has rank m on C. The same holds for g_ν on $\cup_{i<\nu} \bar{W}_i \cap \bar{U}_\nu$. On W_ν we have $\psi \cdot h_\nu = 1$, and hence there $g_\nu = q \circ h_\nu$ also has rank m. Finally,

$$|g_\nu - g_{\nu-1}| \leqslant |q \circ h_\nu - g_{\nu-1}| < \zeta < \epsilon$$

on $\bar{U}_\nu$ and hence everywhere. This concludes the proof of the theorem. $\quad\square$

Underpinning this proof is a general procedure of passing from a local statement – in this case (7.6) – to a global statement.

For an injective immersion one needs more room, as the mapping $S^1 \to \mathbb{R}^2$ in Fig. 48 demonstrates.

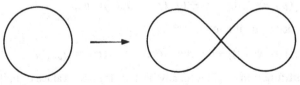

Fig. 48

(7.9) Theorem. *Let $f: M^m \to \mathbb{R}^n$ be a differentiable map and $2m < n$. Let $A \subset M$ be closed and let the restriction of f to a neighbourhood*

U of A be an injective immersion. Then arbitrarily close to f there exists an injective immersion g: M → $\mathbb{R}^n$, so that g|A = f|A.

Proof. As in the previous theorem, we describe the 'nearness' by an everywhere positive function $\delta: M \to \mathbb{R}$. From the previous theorem we may also assume that f is already an immersion. Then by the rank theorem (5.4) f is locally an embedding; we may choose a cover $\{U_\alpha\}$ of M so that for all α, $f|U_\alpha$ is an embedding, and so that $U_\alpha \subset U$ or $U_\alpha \subset M - A$. Then once more using (7.1), we choose a good atlas $\{h_\nu: V_\nu \to K(3)|\nu \in \mathbb{Z}\}$, which is subordinate to this cover, and so numbered that $V_\nu \subset U$ if and only if $\nu \leqslant 0$. Finally, we may choose a bump function ψ for $K(1)$ with support in $K(2)$, and set

$$\psi_\nu := \psi \circ h_\nu: M \to \mathbb{R}$$

Inductively, we construct a sequence of immersions $g_\nu: M \to \mathbb{R}^n$ by

$$g_0 = f,$$
$$g_\nu = g_{\nu-1} + \psi_\nu \cdot b_\nu, \quad b_\nu \in \mathbb{R}^n;$$

where the point b has yet to be specifically chosen.

First, it follows from (7.6) that g_ν has rank m on $h_\nu^{-1}K(2)$ and hence everywhere, provided that b_ν is chosen sufficiently small. Choose b_ν to be also sufficiently small, so that for all x, $|g_\nu(x) - g_{\nu-1}(x)| < 2^{-\nu} \cdot \delta(x)$. Therefore all the functions g_ν, together with $g := \lim_{\nu \to \infty} g_\nu$, remain immersions – they lie in the prescribed neighbourhood of f and agree with f on A. In the choice of b_ν let

$$N^{2m} \subset M \times M$$

be the open subset of points (p, q) with $\psi_\nu(p) \neq \psi_\nu(q)$.

Consider the map

$$N^{2m} \to \mathbb{R}^n, \quad (p, q) \mapsto -(g_{\nu-1}(p) - g_{\nu-1}(q)) \cdot (\psi_\nu(p) - \psi_\nu(q))^{-1}.$$

Because $2m < n$ Sard's theorem implies that the image of this map has measure zero, and we may choose b_ν not to be in this image. Then

$$g_\nu(p) = g_\nu(q) \quad \text{if and only if}$$
$$g_{\nu-1}(p) - g_{\nu-1}(q) = -(\psi_\nu(p) - \psi_\nu(q)) \cdot b_\nu,$$

and so by choice of b_ν if and only if

$$\psi_\nu(p) = \psi_\nu(q) \quad \text{and therefore} \quad g_{\nu-1}(p) = g_{\nu-1}(q).$$

Since the limit function g agrees locally with g_ν for large values of ν, it follows that, if $p \neq q$ and $g(p) = g(q)$, then $g_\nu(p) = g_\nu(q)$ for sufficiently large ν. Hence by downward induction

$$\psi_\nu(p) = \psi_\nu(q) \quad \text{and} \quad g_\nu(p) = g_\nu(q) \quad \text{for all} \quad \nu \geqslant 0.$$

On account of the second condition, in particular, $f(p) = f(q)$, and so p and q cannot lie in the same chart domain V_ν. However, if $p \in W_\nu \subset V_\nu$ and $\nu > 0$, then $\psi_\nu(p) = 1 = \psi_\nu(q)$, implying that $q \in V_\nu$. The remaining possibility is that both p and q lie in a chart domain W_ν with $\nu \leqslant 0$. But in this case, $p, q \in U$ and $f|U = g|U$ is injective. □

An injective immersion is, as we know, in general not yet an embedding, nor is it possible in general to approximate a given map by an embedding (example in the exercises). However, an injective map of locally compact spaces $f: X \to Y$ clearly induces a homeomorphism $f: X \to f(X)$ if it is *proper*, that is, if it may be extended continuously to a map of the one point compactifications $f^0: X^0 \to Y^0$ by mapping the extra point to the extra point. In other words, f is proper if $f^{-1}(K)$ is compact for each compact subset K. In this case, $f(X) \subset Y$ is closed, for $f(X)^0 \subset Y^0$ is compact.

(7.10) Embedding theorem. *An m-dimensional differentiable manifold can be embedded as a closed subset of the Euclidean space $\mathbb{R}^n$, if $2m < n$.*

For this we need:

(7.11) Lemma. *If M is a differentiable manifold and $n > 0$, then there exists a proper differentiable map $M \to \mathbb{R}^n$.*

Proof (of (7.11)). Choose a countable partition of unity $\{\phi_\nu | \nu \in \mathbb{N}\}$ with compact supports Supp (ϕ_ν) and set

$$f = \sum_{\nu=1}^{\infty} \nu \cdot \phi_\nu: M \to \mathbb{R}.$$

If $K \subset \mathbb{R}$ is compact, therefore $K \subset [-n, n]$ for some $n \in \mathbb{N}$, and $f(x) \in K$, then $x \in \bigcup_{\nu=1}^{n}$ Supp (ϕ_ν), and this set is compact. Hence f is proper, and one obtains a proper map $M \to \mathbb{R}^n$ if one chooses f to be the first component. □

Proof (of (7.10)). By (7.11) one can choose a proper map $f: M \to \mathbb{R}^n$, and by (7.9) approximate this by an injective immersion $g: M \to \mathbb{R}^n$, so that $|g - f| \leqslant 1$ and $A = \emptyset$. If $K \subset \mathbb{R}^n$ is compact, then $K \subset K(r)$ for some radius r, hence $g^{-1}(K)$ is closed in the compact set $f^{-1}K(r + 1)$, hence compact. Therefore g is proper, hence an embedding. □

One can improve the results presented here in several ways; as we have already said, one can involve the higher derivatives in the approximations and, by deeper theorems of Whitney and Hirsch, the embedding theorem (7.10) holds also for $n = 2m$. There exists a large literature on embedding and non-embedding theorems. For non-embedding theorems in particular, we lack all the tools here. They substantially depend on the methods of algebraic topology. For example, it is very plausible that there exists no embedding $\mathbb{R}\mathbf{P}^2 \to \mathbb{R}^3$ of the projective plane in the intuitive 'space' of our visual perceptions, but it is an unhappy undertaking to attempt to prove this directly.

(7.12) Remark. From the embedding theorem it follows that a differentiable manifold is homeomorphic to a closed subset of Euclidean space; hence it inherits a *complete metric* from the Euclidean space, *which induces the given topology on the manifold.* This occasionally may simplify arguments from general topology.

(7.13) Exercises

1 Let M be a differentiable manifold and $p \in M$. Show that the map

$$C^\infty(M) \to \mathscr{E}(p), \quad f \mapsto \bar{f}$$

is surjective.

2 Let $A \subset M$ be closed, U an open neighbourhood of A, and f a differentiable map from U into $\mathbb{R}^n$. Show that there exists a differentiable map $g: M \to \mathbb{R}^n$ with $g|A = f$.

3 Construct an injective differentiable map $f: S^1 \to \mathbb{R}^2$, whose image consists of the points $\{x \in \mathbb{R}^2 | \max \{|x_1|, |x_2|\} = 1\}$.

4 Let $f: M \to N$ be a continuous map. Show that f is differentiable if and only if for each $g \in C^\infty(N), g \circ f \in C^\infty(M)$.

5 Show that the ring $\mathscr{E}_n$ possesses divisors of zero.

6 Give an immersion $\mathbb{R} \to \mathbb{R}^2$ (and not just a picture!), which cannot be approximated with proximity? 1 by an embedding.

7 Show that for each n there exists a differentiable map $f: \mathbb{R} \to \mathbb{R}^n$, so that for each $k \in \mathbb{N}$

$$f\{t \in \mathbb{R} \,|\, t \geqslant k\}$$

contains all points, for which all coordinates are rational.

8 Find a function $\delta: \mathbb{R} \to \mathbb{R}, \delta > 0$, and for each $n \in \mathbb{R}$ a differentiable map $f: \mathbb{R} \to \mathbb{R}^n$, so that for no embedding $g: \mathbb{R} \to \mathbb{R}^n$ one has $|g - f| < \delta$.
   Hint: use exercise 7.

9 For a compact manifold M^m it is easy to prove an embedding theorem without regard to the dimension. One can choose a finite good atlas $\{h_\nu | \nu = 1, \ldots, r\}$, a bump function ψ for $K(1)$ with support in $K(2)$, and one sets $\psi_\nu := \psi \circ h_\nu: M \to \mathbb{R}$ and $k_\nu := \psi_\nu \cdot h_\nu: M \to \mathbb{R}^m$ (both maps vanish outside V_ν). Show that the map

$$M \to \prod_{\nu=1}^{r} \mathbb{R}^m \times \prod_{\nu=1}^{r} \mathbb{R}$$

$$p \mapsto (k_1(p), \ldots, k_r(p), \psi_1(p), \ldots, \psi_r(p)),$$

is an embedding, without using anything else from this chapter.

10 Let M^m be a connected non-compact differentiable manifold. Show that there exists a sequence of open subsets $V_\nu \subset M$, so that $V_\nu \cong K(1) \subset \mathbb{R}^m$, $V_\nu \cap V_{\nu+1} \neq \emptyset$, $V_\nu \cap V_\lambda = \emptyset$ if $\lambda \notin \{\nu - 1, \nu, \nu + 1\}$, and $\{V_\nu \mid \nu \in \mathbb{N}\}$ is locally finite, see Fig. 49.

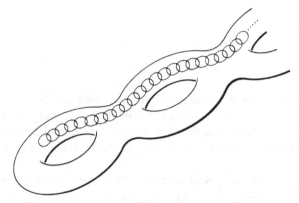

Fig. 49.

11 Show that there exists a closed embedding of the real line in every connected non-compact differentiable manifold.
Hint: use exercise 10.

8
Dynamical systems

The differential topologist sometimes 'pushes' a submanifold aside, 'dents' it somewhere, 'bends' or 'deforms' it, and the handwaving which accompanies such operations all the more undermines the confidence of the observer. He believes the assertions are plausible but that they have not been proven.

We propose to make such 'bending' precise by means of isotopies of embeddings and, in order to be able to construct isotopies, one needs dynamical systems on manifolds. Both for their own importance and for the applications we turn our attention first to these.

(8.1) Definition. Let M be a differentiable manifold. A differentiable map

$$\Phi: \mathbb{R} \times M \to M$$

is called a *dynamical system* or *flow* on M, if for all $x \in M$ and $t, s \in \mathbb{R}$ we have

(i) $\Phi(0, x) = x$,
(ii) $\Phi(t, \Phi(s, x)) = \Phi(t + s, x)$.

The essential content of these two conditions becomes clear if one replaces Φ by a family of maps $M \to M$, parametrised by $\mathbb{R}$. We write

$$\Phi_t: M \to M, \quad x \mapsto \Phi(t, x).$$

Then (i) and (ii) read $\Phi_0 = \mathrm{Id}_M$, and $\Phi_t \circ \Phi_s = \Phi_{t+s}$, so that $\Phi_{-t} = \Phi_t^{-1}$, and one has:

(8.2) Note. A differentiable map $\Phi: \mathbb{R} \times M \to M$ is a dynamical system if and only if the map $t \mapsto \Phi_t$ defines a group homomorphism of the abelian group $(\mathbb{R}, +)$ into the group $\mathrm{Diff}(M)$ of diffeomorphisms of M onto itself. One also says that the group $(\mathbb{R}, +)$ *operates* on M.

Geometrically, one takes a quite different position if one considers the flow $\Phi: \mathbb{R} \times M \to M$ as a family of curves $\mathbb{R} \to M$ parametrised by M.

74

(8.3) Definition. If Φ: $\mathbb{R} \times M \to M$ is a flow, and $x \in M$, the curve

$$\alpha_x: \mathbb{R} \to M, \quad t \mapsto \Phi_t(x)$$

is called the *flow line* or *integral curve* of x. The image $\alpha_x(\mathbb{R})$ of the flow line is called the *orbit* of x, see Fig. 50.

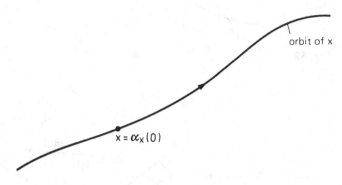

orbit of x

$x = \alpha_x(0)$

Fig. 50

(8.4) Remark. If a flow is given on a manifold, then exactly one orbit passes through each point p of M.

Proof. The relation $x \sim y \Leftrightarrow x = \Phi_t(y)$ for some t is an equivalence relation for points of M, as one can easily check. The orbits are the equivalence classes. $\qquad\square$

In order to obtain an idea of the geometric mechanism of a flow, one does not usually consider the single diffeomorphism Φ_t, but one tries to give an overall picture of the behaviour of all the orbits. There are three types of orbit:

(8.5) Remark. A flow line α_x: $\mathbb{R} \to M$ of a flow is either an injective immersion (Fig. 51), or a *periodic* immersion (Fig. 52), that is, α_x is an immersion and there exists some $p > 0$ with $\alpha_x(t+p) = \alpha_x(t)$ for all t; or α_x is constant, $\alpha_x(t) = x$ for all t. In the last case x is called a *fixed point* of the flow.

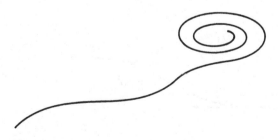

Fig. 51

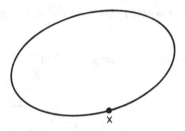

Fig. 52

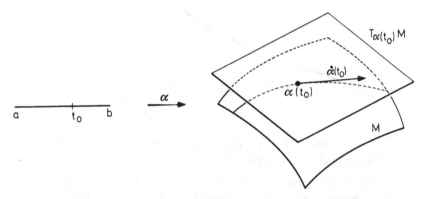

Fig. 53

Proof. If $\alpha: (a, b) \to M$ is a differentiable curve in M and $t_0 \in (a, b)$, then we write $\dot{\alpha}(t_0) \in T_{\alpha(t_0)}M$ for the *velocity vector of the curve at the point* t_0. Thus $\dot{\alpha}(t_0)$ as a derivation is given by $\dot{\alpha}(t_0)(f) := (\mathrm{d}/\mathrm{d}t)f\alpha(t_0)$, see Fig. 53. Now for a flow line α_x (see Fig. 54) we have $\dot{\alpha}_x(t_0) = T(\Phi_{t_0})(\dot{\alpha}_x(0))$, for $\alpha_x(t + t_0) = (\Phi_{t_0} \circ \alpha_x)(t)$. Since Φ_{t_0} is a diffeomorphism, either $\dot{\alpha}_x(t) \neq 0$ for all t, that is, the flow line is an immersion (non-singular curve) or, $\dot{\alpha}_x(t) = 0$ for all t, that is, α_x is constant. If α_x is not injective, hence $\alpha_x(t_0) = \alpha_x(t_1)$ for specific values $t_0 < t_1$, then $\Phi_{t_0}(x) = \Phi_{t_1}(x)$, hence also $\Phi_t\Phi_{t_0}(x) = \Phi_t\Phi_{t_1}(x)$ for all t. It follows that $\Phi_t(x) = \Phi_{t+(t_1-t_0)}(x)$, that is, $\alpha_x(t) = \alpha_x(t + (t_1 - t_0))$ for all t. $\qquad\square$

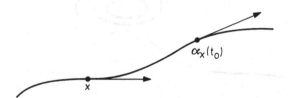

Fig. 54

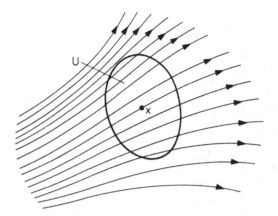

Fig. 55

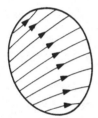

Fig. 56

If we have a flow on M and U is an open subset of M, then we see from Fig. 55 that, in general, it is not the case that flow lines of points in U lie entirely in U. On account of continuity however, if $x \in U$ the flow line α_x must belong to U for some small interval (a_x, b_x) about $0 \in \mathbb{R}$, see Fig. 56. This situation leads us to the definition of the concept of a 'local flow':

(8.6) Definition. Let M be a differentiable manifold. By a *local flow* Φ on M we understand a differentiable map

$$\Phi : A \to M$$

from an open subset $A \subset \mathbb{R} \times M$, containing $0 \times M$, to M, so that for each $x \in M$ the intersection $A \cap (\mathbb{R} \times \{x\})$ is connected, see Fig. 57, and so that

(i) $\Phi(0, x) = x$
(ii) $\Phi(t, \Phi(s, x)) = \Phi(t + s, x)$

for all t, s, x for which both sides are defined.

A local flow with $A = \mathbb{R} \times M$ is clearly a flow (*global flow*).

(8.7) Notation. If $\Phi : A \to M$ is a local flow on M, then we shall

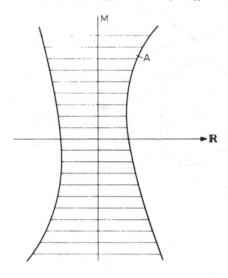

Fig. 57

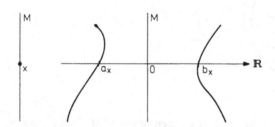

Fig. 58

denote the domain of definition of the flow line α_x:

$$t \mapsto \Phi(t, x)$$

by (a_x, b_x), see Fig. 58.

Note that for a local flow, one can, in general, no longer talk of the diffeomorphism Φ_t, since for fixed $t \neq 0$, $x \mapsto \Phi(t, x)$ is not necessarily defined on all of M, see Fig. 59.

   (8.8) Definition. If Φ is a (local or global) flow on M, then the vector field

$$\Phi: M \to TM, \quad x \mapsto \dot\alpha_x(0)$$

is called the *velocity field* of the flow, see Fig. 60.

   (8.9) Remark. For all flow lines and for all $t \in (a_x, b_x)$, $\dot\alpha_x(t) = \Phi(\alpha_x(t))$, see Fig. 61.

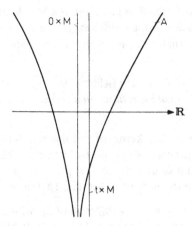

Fig. 59

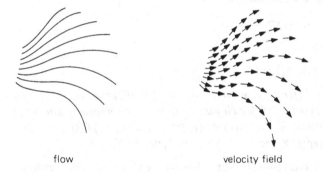

flow   velocity field

Fig. 60

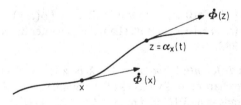

Fig. 61

Proof. This follows from the definition for $t = 0$. For $z = \alpha_x(t)$, we have $\alpha_z(s) = \alpha_x(s + t)$, provided that both sides are defined (in any case in some neighbourhood of $s = 0$), hence $\dot{\alpha}_z(0) = \dot{\alpha}_x(t)$. □

Often in geometric considerations, one needs flows which 'do' something or other, that is, have preassigned properties. It would be highly inconvenient to always have to explicitly construct such flows as maps $\mathbb{R} \times M \to M$ or $A \to M$. What really makes flows useable is the result that a flow is completely determined by its velocity field and that, to a prescribed velocity field, there actually exists a flow.

(8.10) Integrability theorem for vector fields. *Every vector field is the velocity field of exactly one maximal local flow; on a compact manifold even of a global one.*

Proof. The essential mathematical kernel of this theorem is the theorem on the existence and uniqueness of solutions to first order ordinary differential equations, which we want to quote here. Our problem consists then only in the translation into the language of manifolds. Therefore:

Quotations from the theory of ordinary differential equations: Let $\Omega \subset \mathbb{R}^n$ be an open subset and $f : \Omega \to \mathbb{R}^n$ a differentiable (C^∞) map. Then we have

(a) Uniqueness theorem. *If*

$$\alpha : (a_0, a_1) \to \Omega$$

and

$$\beta : (b_0, b_1) \to \Omega$$

are differentiable curves with $\alpha(0) = \beta(0) = x$, and $\dot{\alpha}(t) = f(\alpha(t))$, $\dot{\beta}(t) = f(\beta(t))$ for all values of t in the appropriate domain of definition, then $\alpha(t) = \beta(t)$ for all $t \in (a_0, a_1) \cap (b_0, b_1)$, (Lang [2], chapter 8, section 1, theorem 3, p. 375.)

(b) Existence theorem. *For each $x \in \Omega$ there exists an open neighbourhood $W \subset \Omega$, some $\epsilon > 0$ and a differentiable (C^∞) map*

$$\phi : (-\epsilon, \epsilon) \times W \to \Omega$$

with the property that $\phi(0, x) = x$ for all $x \in W$, and $\dot{\phi}(t, x) = f(\phi(t, x))$ for all $(t, x) \in (-\epsilon, \epsilon) \times W$. (Lang [2], chapter 8, section 4, theorem 7, p. 388.)

Connection with differential topology. Let X be a vector field on M and (h, U) a differentiable chart of M. By means of the bundle chart of TM associated to (h, U), we transplant $X \,|\, U$ to a map $f : U' \to TU' = U' \times \mathbb{R}^n \to \mathbb{R}^n$ of U' into $\mathbb{R}^n$, namely, $f(h(x)) := T_x h(X(x))$, see Fig. 62. Here $T_{h(x)} U' \cong \mathbb{R}^n$ in the usual way. Then for curves $\alpha : (a, b) \to U$ we have

$$\dot{\alpha}(t) = X(\alpha(t)) \Leftrightarrow (h \cdot \alpha)^{\cdot}(t) = f(h \cdot \alpha(t)), \text{ see Fig. 63.}$$

We now want to call a curve $\alpha : (a, b) \to M$ a solution curve for X, if $\dot{\alpha}(t) = X(\alpha(t))$ everywhere. Then the considerations above show that, for each $x \in M$, there is exactly one maximal solution curve $\alpha_x : (a_x, b_x) \to M$ with

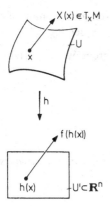

Fig. 62

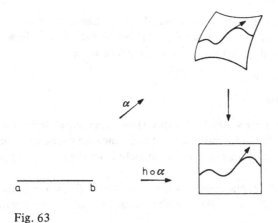

Fig. 63

$\alpha_x(0) = x$. The existence of a solution curve with $\alpha(0) = x$ follows (using a chart about x) from the existence theorem for ordinary differential equations, and any two solution curves agree on the intersection of their intervals of definition. This follows since the set of t, where the two solutions agree, is closed by continuity, but it is also open, as one sees by applying the uniqueness theorem in the image of a chart about $y \in M$. Therefore, the

Fig. 64

uniquely determined maximal solution curve is given on the union of all intervals of definition of all solution curves with $\alpha(0) = x$, see Fig. 64.

Now let us turn to the proof proper of the theorem. We first establish the following assertion:

(8.11) Assertion. *The set*

$$A := \bigcup_{x \in M} (a_x, b_x) \times x \ ,$$

determined by the domains of definition of maximal solution curves is open in $\mathbb{R} \times M$, *and the map*

$$\Phi : A \to M$$

given by the solution curves is a maximal local flow with the given vector field as velocity field.

To prove this it is enough to show that A is open and Φ is differentiable, since the conditions $\Phi(0, x) = x$ and $\Phi(t, \Phi(s, x)) = \Phi(t + s, x)$ follow simply from the fact that $\Phi|(a_x, b_x) \times x$ is a solution curve. Both

$$t \mapsto \Phi(t + s, x)$$

and

$$t \mapsto \Phi(t, \Phi(s, x))$$

(where we allow all t for which both expressions make sense) define maximal solution curves for the initial value $\Phi(s, x)$ and hence are necessarily identical. The maximality of the flow follows immediately from the maximality of the solution curves.

Now for each $x \in M$, one considers the interval $J_x \subset \mathbb{R}_+$, which consists of those $t \geq 0$, for which A contains a neighbourhood of $[0, t] \times x$, on which Φ is differentiable.

Then we have to show that $J_x = [0, b_x)$ and the corresponding result for $t \leq 0$. By definition, J_x is open and it is enough to show that J_x is non-empty and closed in $[0, b_x)$. Both follow from the local existence theorem:

For a point $p \in M$ we find a neighbourhood W of p in M, an $\epsilon > 0$ and a differentiable map

$$\phi : (-2\epsilon, 2\epsilon) \times W \to M,$$

so that $\phi|(-2\epsilon, 2\epsilon) \times q$ is a solution curve for the initial value $q \in W$. From this follows, first of all, that A contains a neighbourhood of $0 \times M$, on which Φ is differentiable, for, given the uniqueness of the solution curves, we must have $\Phi|(-2\epsilon, 2\epsilon) \times W = \phi$. Hence J_x is non-empty. If $\tau \in \bar{J}_x$ (closure in $[0, b_x)$!) and $\Phi_\tau(x) = p$ then, by definition of J_x, we have a set $[0, \tau - \epsilon] \times U$ in A, in whose neighbourhood Φ is defined and differentiable. Here, U is a neighbourhood of x in M, and ϵ is chosen as above for the point p with $\tau - 2\epsilon > 0$. If one now defines the neighbourhood U' of x in M by

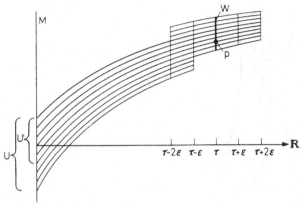

Fig. 65

$$U' = \Phi_{\tau-\epsilon}^{-1}\phi_{-\epsilon}(W),$$

with W the neighbourhood of p chosen above, see Fig. 65, then Φ is defined and differentiable in a neighbourhood of $[0, \tau + \epsilon] \times U'$, hence particularly in a neighbourhood of $[0, \tau] \times x$. Note that the differentiable map

$$(\tau - 2\epsilon, \tau + 2\epsilon) \times U' \to M$$

$$(t, u) \mapsto \phi(t - \tau, (\tau, u))$$

correctly extends the solution curves given by Φ on $U' \times [0, \tau - \epsilon]$ because of the uniqueness theorem, see Fig. 66. Therefore $\tau \in J_x$, which is what we had to show.

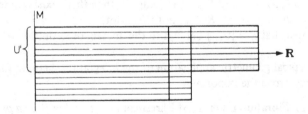

Fig. 66

In this way, we have associated a maximal local flow to the preassigned velocity field. That this is the only maximal one follows immediately from (8.11), for each flow with the same velocity field must be a restriction of Φ since its flow lines are solution curves of the field and Φ has the maximal solution curves as flow lines. Thus, the uniqueness part of the integrability theorem for vector fields is also proved, and it only remains to show that the *maximal flow of a velocity field given on a compact manifold is global.*

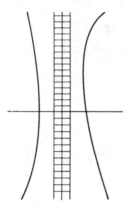

Fig. 67

If M is compact, then for some $\epsilon > 0, A$ contains a subset of the form $(-\epsilon, \epsilon) \times M$, see Fig. 67. Then $(-2\epsilon, 2\epsilon) \times M$ must also be contained in A, for one can extend the flow defined on $(-\epsilon, \epsilon) \times M$ to $(-2\epsilon, 2\epsilon) \times M$ by setting

$$\Phi(t, x) := \Phi\left(\frac{t}{2}, \Phi\left(\frac{t}{2}, x\right)\right).$$

Since $\Phi: A \to M$ is maximal, it follows that $(-2\epsilon, 2\epsilon) \times M \subset A$. Clearly, therefore, $\mathbb{R} \times M = A$, which concludes the proof. $\qquad\square$

A generalisation of this last part of the theorem actually holds: a maximal solution curve, which is not defined for all time, eventually leaves each compact set. This means, if $\alpha: (a_x, b_x) \to M$ is a maximal solution curve of a vector field on $M, b_x < \infty$ and $K \subset M$ is compact, then there exists some $\epsilon > 0$, so that $\alpha(b_x - \epsilon, b_x) \cap K = \emptyset$. For the proof, one needs only to choose ϵ so small that $K \times [0, \epsilon]$ belongs to the domain of definition A of the local flow.

As a first typical geometric application of the integration theorem for vector fields we prove the important

(8.12) Fibration theorem of Ehresmann. *Let* $f: E \to M$ *be a proper submersion of differentiable manifolds, then* f *is a locally trivial fibration, that is, if* $p \in M$ *and* $F = f^{-1}(p)$ *the fibre of* p, *then there exists a neighbourhood* U *of* p *in* M *and a diffeomorphism* $\phi: U \times F \to f^{-1}U$, *so that the following diagram is commutative:*

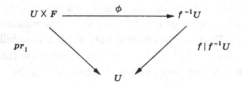

Proof. The assertion is local relative to M, so that we may replace E, M and f by $f^{-1}U$, U and the restriction of f, and thus, w.l.o.g., assume that $M = \mathbb{R}^n$ and $p = 0$. In this case we have the basic vector fields $\partial/\partial x_\nu$, and we can lift these to E, obtaining vector fields $v_1, \ldots, v_n$ on E, so that for all $x \in E$

$$T_x f(v_\nu(x)) = \partial/\partial x_\nu.$$

Locally, about a point $x \in E$, such fields are easy to find because, by the rank theorem, f is transformable to the form $pr_1 \colon U \times V \to U$, and one obtains the v_ν on all of E by glueing together the locally chosen fields by means of a partition of unity.

Now by (8.8), (8.10), the vector fields v_ν determine local flows Φ^ν on E, and in order to prove the theorem we put

$$F = f^{-1}(0), \phi(u,x) = \Phi^1_{u_1} \circ \ldots \circ \Phi^n_{u_n}(x)$$

for $x \in F$ and $u = (u_1, \ldots, u_n) \in M = U = \mathbb{R}^n$.

It is perhaps not immediately clear that the map $\phi(u, x)$, defined in this way, actually exists for all $u \in \mathbb{R}^n$, but in any case, so long as the local flows exist,

for,
$$u = pr_1(u,x) = f \circ \phi(u,x),$$
$$f \circ \Phi^\nu_{u_\nu}(y) = f \circ \Phi^\nu_0(y) + u_\nu e_\nu,$$

where $e_\nu \in \mathbb{R}^n$ is the νth unit vector. Namely, the equation holds for $u_\nu = 0$, and the agreement of the derivatives according to u_ν is assured, because v_ν lifts the field $\partial/\partial x_\nu$.

Then, however, it also follows that all the flow maps in the definition of ϕ exist, because for $|u| \leqslant K$ the flow lines remain inside the compact set $f^{-1}\{u \in \mathbb{R}^n | u| \leqslant K\}$. Here we use the assumption that f is proper. Finally, one obtains the inverse map $\phi^{-1} \colon E \to U \times F$ for $U = \mathbb{R}^n$ by setting $f(y) = u$ and

$$\phi^{-1}(y) = (u, \Phi^n_{-u_n} \circ \ldots \circ \Phi^1_{-u_1}(y)). \qquad \square$$

The assumption that f is proper is essential; if, for example, we remove a point from E, the restriction of f is still a submersion but, in general, it is no longer a fibration.

(8.13) Exercises

1 Show that for each $n \geqslant 0$ there is a flow on S^1 with exactly n fixed points.
2 Show that for each vector field X on M there is an everywhere positive function $\epsilon \colon M \to \mathbb{R}$, such that ϵX is globally integrable.
3 Show that each bounded vector field defined on $\mathbb{R}^n$ is globally integrable.
4 Let $G \subset \mathbb{R}$ be a closed subset and subgroup of $(\mathbb{R}, +)$. Show that either $G = 0$, or $G \cong \mathbb{Z}$, or $G = \mathbb{R}$.
   Let $\alpha_x \colon \mathbb{R} \to M$ be a flow line of a dynamical system; show that

$G := \{t \in \mathbb{R} \mid \alpha_x(t) = x\}$ is a closed subgroup of $(\mathbb{R}, +)$, and that the following hold:

α_x is an immersion iff $G \neq \mathbb{R}$.

α_x is periodic iff $G \cong \mathbb{Z}$. The smallest period is then a generator for G.

If α_x is periodic, then $\alpha_x(\mathbb{R}) \subset M$ is a submanifold, diffeomorphic to a circle.

5 Let M be a compact manifold of dimension $\geqslant 2$. Show that there exists an injective immersion $\mathbb{R} \to M$, whose image is not a flow line of a flow on M.

6 Show that every submanifold of M diffeomorphic to S^1 arises as the orbit of a global flow on M.
Hint: partitions of unity.

7 An open set $U \subset \mathbb{R}^n$ is called *star-shaped* for $p \in U$ if, for each $x \in U$, the line segment joining x to p lies entirely in U. Show that a star-shaped subset of $\mathbb{R}^n$ is diffeomorphic to $\mathbb{R}^n$.
Hint: construct a diffeomorphism which maps the orbits of the vector field $X(x) = x - p$ on $\mathbb{R}^n$ onto the orbits of a vector field $\epsilon \cdot X$ on U, with ϵ as in exercise 2, see Fig. 68.

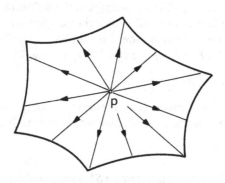

Fig. 68

8 Give an example of a fixed point free flow on S^{2n-1}.
Hint: $S^{2n-1} \subset \mathbb{C}^n$.

9 Define a flow on S^2, which has exactly two fixed points, and exactly one closed orbit.

10 Give an example of a flow on the projective plane $\mathbb{R}P^2$, which has exactly one fixed point and otherwise only closed orbits.

11 For each $\lambda \in [0, 1]$ let a flow $\Phi^{(\lambda)} \colon \mathbb{R} \times S^1 \to S^1$ be given, so that the associated map $[0, 1] \times \mathbb{R} \times S^1 \to S^1$ is differentiable, and so that $\Phi^{(1)}$ is the reversed flow for $\Phi^{(0)}$, that is, $\Phi^{(1)}(t, x) = \Phi^{(0)}(-t, x)$. Show that each point $x \in S^1$ is a fixed point of $\Phi^{(\lambda)}$ for some λ, see Fig. 69.

Fig. 69

12 Show that if X is a vector field on S^2, which is nowhere tangential to the 'equator' $S^1 = S^2 \cap (\mathbb{R}^2 \times 0) \subset \mathbb{R}^3$, then each flow line meets the equator at most once.

13 Show that on the torus $S^1 \times S^1$ there exists a vector field for which no orbit of the associated flow is a submanifold of $S^1 \times S^1$.
   Hint: $S^1 \times S^1 = \mathbb{R} \times \mathbb{R} / \mathbb{Z} \times \mathbb{Z}$. Consider a specific constant vector field on $\mathbb{R}^2$.

14 Show that on each non-compact connected manifold there exists a vector field which is not globally integrable.
   Hint: apply exercise 11 of Chapter 7.

15 Let $g \colon \mathbb{R} \to \mathbb{R}_+$ be continuous, $\lim\limits_{|x| \to \infty} g(x) = 0$, and $A = \{(t, x) \in \mathbb{R}^2 \mid t < g(x)\}$. Show that there is a maximal local flow on $\mathbb{R}$, which is defined on A but not on all of $\mathbb{R} \times \mathbb{R}$.

9
Isotopy of embeddings

For the intuitive, as well as the formal, understanding of the theory of differentiable manifolds, it is important to know the extent to which submanifolds can be 'moved'.

(9.1) Definition. Let $f: M \to N$ be an embedding. A differentiable map $h: [0, 1] \times M \to N$ is called an *isotopy* of f if $h_0 = f$ and each of the maps

$$h_t: M \to N \quad x \mapsto h(t, x)$$

is an embedding; h is called an isotopy between h_0 and h_1, and h_0 and h_1 are called isotopic embeddings, see Fig. 70.

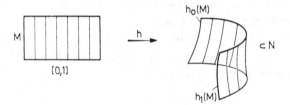

Fig. 70

At the 'boundary points', for example $(0, x)$ by 'differentiable' we mean that there exists some neighbourhood $\tilde{U}$ of $(0, x)$ in $\mathbb{R} \times M$ and a differentiable map $\tilde{h}: \tilde{U} \to N$, which agrees with h on $\tilde{U} \cap ([0, 1] \times M)$, see Fig. 71. Although this is the way one thinks and speaks about isotopy, it is often technically more convenient to use a modified (but equivalent) definition. For example,

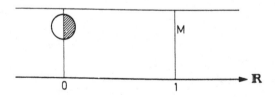

Fig. 71

88

without further assumptions, the definition above does not imply that isotopy between embeddings is a transitive relation. Thus, when we naively stick together isotopies h between f and f', and k between f' and f'':

$$(t, x) \mapsto \begin{cases} h(2t, x) & \text{for } 0 \leqslant t \leqslant \frac{1}{2} \\ h(2t - 1, x) & \text{for } \frac{1}{2} \leqslant t \leqslant 1, \end{cases}$$

illustrated by Fig. 72; then, because $h_1 = k_0 = f'$, this map is certainly continuous but, in general, not differentiable.

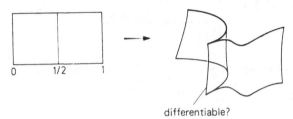

differentiable?

Fig. 72

(9.2) Definition. A differentiable map $h : \mathbb{R} \times M \to N$ will be called a *technical* isotopy, if each h_t is an embedding and also for some $\epsilon > 0$.

$$h_t = \begin{cases} h_0 & \text{for } t \leqslant \epsilon \\ h_1 & \text{for } t \geqslant 1 - \epsilon \end{cases}$$

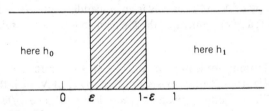

here h_0 here h_1

0   ε $1-\varepsilon$ 1

Fig. 73

It is clear that one can easily join such technical isotopies together, as in Fig. 74, and that the combined isotopy is again technical. If h is a technical isotopy between h_0 and h_1, then clearly $h \,|\, [0, 1] \times M$ is an isotopy between h_0 and h_1. Conversely, given an isotopy $h : [0, 1] \times M \to N$ between h_0 and h_1, and a C^∞-function $\phi : \mathbb{R} \to [0, 1]$ of the kind illustrated by Fig. 75 (compare Chapter 7), then the map

$$\mathbb{R} \times M \to N$$

defined by

$$(t, x) \mapsto h(\phi(t), x)$$

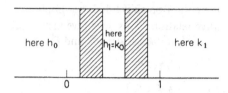

Fig. 74

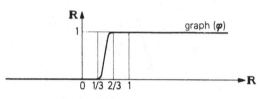

Fig. 75

is a technical isotopy between h_0 and h_1. In particular therefore, 'isotopic' is an equivalence relation.

(9.3) Definition. By a *diffeotopy* of a manifold N we understand a differentiable map

$$H: [0, 1] \times N \to N,$$

such that $H_0 = \text{Id}_N$ and each $H_t: N \to N$ is a diffeomorphism.

If H is a diffeotopy of N and $f: M \to N$ is an embedding, then $h_t := H_t \circ f$ gives an isotopy of f; any movement of the big manifold carries all submanifolds with it.

(9.4) Definition. An isotopy $h: [0, 1] \times M \to N$ is said to be *embeddable* in a diffeotopy if there exists a diffeotopy H of N, such that for all t, $h_t = H_t \circ h_0$. The embeddings h_0 and h_1 are then said to be *diffeotopic in N.*

Two diffeotopic embeddings h_0 and h_1 are, in particular, clearly equivalent in the sense that there exists a diffeomorphism (here equal to H_1) from N to itself, so that is commutative, something which for merely isotopic embeddings need not be the case. One is frequently in the situation where one has isotopy and would like to have diffeotopy. The following theorem, which constitutes the main content of the present chapter, shows that, under certain conditions, this wish can be fulfilled.

(9.5) **Theorem (R. Thom 1957).** *If h is a (technical) isotopy of embeddings of M in N, which holds fixed all points outside a compact subset M_0 of M, then one can embed h in a (technical) diffeotopy of N, and indeed even in one which holds fixed all points outside a compact subset N_0 of N.*

Although the theorem is valid for arbitrary isotopies, we shall only prove the weaker result for technical isotopies. This is enough for all applications – in particular, one can conclude from the existence of an isotopy the existence of an embedded isotopy.

Proof of the theorem. Let $h \colon \mathbb{R} \times M \to N$ be a technical isotopy which holds fixed every point outside the compact set $M_0 \subset M$. We can choose a compact neighbourhood N_0 of $h([0, 1] \times M_0)$, as in Fig. 76. We want to construct a technical diffeotopy $H \colon \mathbb{R} \times N \to N$, which holds fixed all points outside N_0 and which has the required property that $h_t = H_t \cdot h_0$, see Figs. 77 and 78. To this end we consider first the map

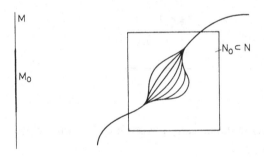

Fig. 76

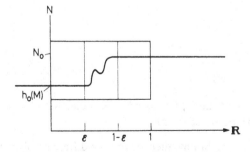

Fig. 77

$F \colon \mathbb{R} \times M \to \mathbb{R} \times N$

$(t, x) \mapsto (t, h_t(x)).$

In order to embed h in a diffeotopy we try to define a global flow on $\mathbb{R} \times N$,

$\Phi \colon \underset{t}{\mathbb{R}} \times (\underset{\tau}{\mathbb{R} \times N}) \to \mathbb{R} \times N,$

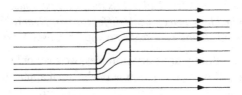

Fig. 78

on which we impose the following conditions:

Condition (i). Φ_t should map $\tau \times N$ onto $(\tau + t) \times N$, see Fig. 79.

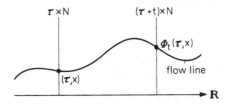

Fig. 79

Condition (ii). Φ should carry the isotopy along with it, which

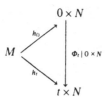

is commutative for all t; and finally

Condition (iii). Outside $[\epsilon, 1 - \epsilon] \times N_0$ the projection of an arbitrary flow line on N should be locally constant.

If Φ fulfills these three conditions, then clearly the differentiable map $H \colon \mathbb{R} \times N \to N$, defined by the diagram

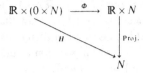

has the properties required by the theorem. Because Φ is a flow, $H_0 = \mathrm{Id}_N$; condition (i) implies that each $H_t\colon N \to N$ is a diffeomorphism ($\Phi_{-t}|t \times N$ takes care of the inverse); because of (ii) $h_t = H_t \circ h_0$ and because of (iii) H is a technical diffeotopy which holds fixed all points outside N_0.

Since every flow is determined by its velocity vector field, we may formulate the conditions (i)–(iii) as conditions on $\dot{\Phi}$.

Assertion 1 in the proof. If Φ is a global flow on $\mathbb{R} \times N$, then the conditions (i)–(iii) are equivalent to the conditions (i′)–(iii′) on $X := \dot{\Phi}$:

(i′) The $\mathbb{R}$-component of X, that is, the image of X under the differential of the projection $\mathbb{R} \times N \to \mathbb{R}$ is equal everywhere to the 'unit tangent vector' $\partial/\partial t$.

(ii′) On $F(\mathbb{R} \times M)$ the field X is given by $T_{(t,x)}F(\partial/\partial t) = X(F(t,x))$, see Fig. 80. This means that the curves $\mathbb{R} \to \mathbb{R} \times N$, $(t \mapsto (t, h_t(x)))$ given by the isotopy are solution curves of X, hence flow lines of Φ, and this again means precisely that the isotopy is 'carried' as in (ii) above.

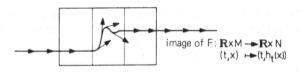

image of $F\colon$ **R**×**M** $\longrightarrow$ **R**× **N**
$(t,x) \longmapsto (t, h_t(x))$

Fig. 80

(iii′) Outside $[\epsilon, 1 - \epsilon] \times N_0$ the field X equals $\partial/\partial t$.

Assertion 2 in the proof. If a vector field X on $\mathbb{R} \times N$ has the properties (i′)–(iii′), then it is the velocity field of a global flow Φ, since $[0, 1] \times N_0$ is compact and the maximal solution curves for initial points outside $[0, 1] \times N_0$ have at least $(-\epsilon, \epsilon)$ in the domain of definition. Hence $(-\delta, \delta) \times (\mathbb{R} \times N) \subset A$ for some $\delta > 0$, hence also $(-2\delta, 2\delta) \times (\mathbb{R} \times N) \subset A$, etc.

We therefore obtain, as an intermediate result: the theorem is proved once we can find a vector field X on $\mathbb{R} \times N$ with the properties (i′)–(iii′).

First of all, we remark that the conditions (i′)–(iii′) for the section $X\colon \mathbb{R} \times N \to T(\mathbb{R} \times N)$ are conditions on the individual vectors $X(t, x)$, and that if the conditions are satisfied for v and w out of $T_{(t,x)}(\mathbb{R} \times N)$, then they are also satisfied for all $\lambda v + (1 - \lambda)w$. Hence it is enough to show that such a vector field exists locally about each point because we can then construct the required vector field on all of $\mathbb{R} \times N$ by means of a partition of unity.

If, for each point outside the compact (and hence closed) subset $F([\epsilon, 1 - \epsilon] \times M_0) \subset \mathbb{R} \times N$, one defines X as $\partial/\partial t$, then one has already solved the local construction problem for all points in $\mathbb{R} \times N - F([\epsilon, 1 - \epsilon] \times M_0)$ (Fig. 81). We therefore consider a point $q_0 = F(t_0, p_0)$ with (t_0, p_0) belonging to $[\epsilon, 1 - \epsilon] \times M_0$. We want a neighbourhood U of q_0 in $\mathbb{R} \times N$ and a vector field X_0 on U with the properties (i')–(iii').

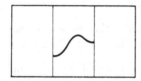

Fig. 81

First, we choose local coordinates about the point q_0 in $t_0 \times N$, with respect to which $h_{t_0}(M)$ is given by $x_{k+1} = \ldots = x_n = 0$. This is possible because h_{t_0} is an embedding, see Fig. 82. With respect to these coordinates and on a sufficiently small neighbourhood of $(t_0, p_0, 0)$ in $\mathbb{R} \times M \times \mathbb{R}^{n-k}$, the map

$$(t, p, x_{k+1}, \ldots, x_n) \mapsto F(t, p) + (0, 0, \ldots, 0, x_{k+1}, \ldots, x_n)$$

is a differentiable map into $\mathbb{R} \times N$, which has maximal rank at the point $(t_0, p_0, 0)$ and, therefore, is a local diffeomorphism. We may choose $\delta > 0$ and a small neighbourhood V of p_0 in M, so that on

$$W := (t_0 - \delta, t_0 + \delta) \times V \times \{x \in \mathbb{R}^{n-k} \mid |x| < \delta\},$$

this map, which we now want to label $\hat{F}$, defines a diffeomorphism

$$\hat{F}: W \to \hat{F}(W) =: U.$$

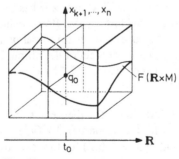

Fig. 82

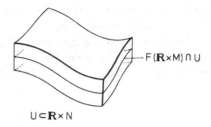

Fig. 83

We may also choose W so small that the projection from U on N remains inside N_0 and that, apart from the points $\hat{F}(t, p, 0)$, no other points of $F(\mathbb{R} \times M)$ lie in U, see Fig. 83. (If this last condition could not be fulfilled, then there would exist a sequence $(t_i, p_i)_{i \in \mathbb{N}}$ with $t_i \mapsto t_0, p_i \in M - V$ and $F(t_i, p_i) \to q_0$. It would be impossible for infinitely many points p_i to belong to the compact set $M_0 - V$, because these would have an accumulation point $\bar{p} \in M - V$, for which $F(t_0, \bar{p}) = q_0$, contradicting the injectivity of h_{t_0}. Hence only finitely many points p_i belong to M_0, and the sequence (t_i, p_i) belongs ultimately to $\mathbb{R} \times (M - M_0)$. However, there h is, by assumption, independent of t, hence not only does $F(t_i, p_i) \to q_0$, but also $F(t_0, p_i) \to q_0$, but then h_{t_0} could not be an embedding: a contradiction.) Next, one carries the vector field $\partial/\partial t$ on W over to a vector field X_0 on U by means of $\hat{F}$:

$$X_0(u) = T_u(\hat{F}^{-1})^{-1}(\partial/\partial t),$$

and obtains thereby a local vector field near q_0 with the properties (i$'$), (ii$'$), (iii$'$).

This closes the gap in the construction, and proves the 'isotopy theorem' (9.5). □

(9.6) Exercises

1. Let M be a connected manifold with dim $(M) \geqslant 2$. Let $x_1, \ldots, x_k$ be distinct points of M, and let $y_1, \ldots, y_k$ also be distinct points of M. Show that there exists a diffeomorphism $\phi: M \to M$ with $\phi(x_i) = y_i$ $(i = 1, 2, \ldots, k)$.

2. Let M be a closed submanifold of the connected manifold N, codim $M \geqslant 2$, and $p, q \in N - M$. Show that there exists a diffeomorphism of N to itself, which is the identity on M, and which maps p to q.

3. If $\Phi: \mathbb{R} \times M \to M$ is a flow, then $\Phi|[0, 1] \times M$ is of course a diffeotopy. Give an example of a diffeotopy which is *not* the restriction of a flow.

4. Let $K \subset \mathbb{R}^n$ be compact and $U \subset \mathbb{R}^n$ open and non-empty. Construct a globally integrable vector field on $\mathbb{R}^n$ so that $\Phi_1(K) \subset U$.

5 Show that in a differentiable vector bundle every differentiable section is an embedding isotopic to the zero-section.

6 Consider the embedding $S^1 + S^1 \to \mathbb{C}$, which is the usual inclusion on the first factor and which is given on the second by $x \mapsto 2x$, see Fig. 84. Define an isotopy of this embedding $h_\tau \colon S^1 + S^1 \to \mathbb{C}$ by

$$e^{2\pi i t} \longmapsto e^{2\pi i (t+\tau)},$$

$$e^{2\pi i s} \longmapsto 2e^{2\pi i (s-\tau)}$$

for $0 \leqslant \tau \leqslant 1$, and embed it in a diffeotopy.

Fig. 84

7 Show that the antipodal map

$$S^n \to S^n \quad x \mapsto -x$$

is isotopic to the identity if and only if n is odd.

8 Construct an embedding $f \colon \mathbb{R} \to \mathbb{R}$ with $f(\mathbb{R}) = (0, 1)$.

9 Give an isotopy of the embedding

$$(0, 1) \to \mathbb{R}^2 \quad t \mapsto (t, 0),$$

which cannot be embedded in a diffeotopy of $\mathbb{R}^2$.

10 Show that any two orientation preserving embeddings $\mathbb{R} \to \mathbb{R}$ are isotopic.

11 Let $n > m$. Show that two arbitrary embeddings $\mathbb{R}^m \to \mathbb{R}^n$ are isotopic.

12 Give two orientation preserving but not diffeotopic embeddings $\mathbb{R} \to \mathbb{R}$.

13 Show that the embeddings in Fig. 85,

$$S^1 \subset \mathbb{R}^2 - \{0\},$$

and

$$S^1 \to \mathbb{R}^2 - \{0\}$$

$$x \mapsto x + (2, 0)$$

are not isotopic in $\mathbb{R}^2 - \{0\}$.

Hint: use complex variable theory.

Fig. 85

14 Find an isotopy $h: \mathbb{R} \times M \to N$ which is such that the map

   $\mathbb{R} \times M \to \mathbb{R} \times N$ $(t,x) \mapsto (t, h(t,x))$

   fails to be an embedding.
   Hint: try $M = \mathbb{R}, N = \mathbb{R}^2$.

10
Connected sums

It is intuitively clear how one can combine two connected manifolds M_1 and M_2 into a third connected manifold $M_1 \# M_2$ (Fig. 86). We treat this process in this section as an application of the isotopy theorem (9.5), because it is the isotopy theorem which shows why the result $M_1 \# M_2$ is essentially well defined, hence independent of the technicalities of the combination.

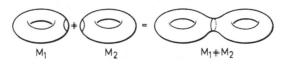

$$M_1 \qquad M_2 \qquad M_1 \# M_2$$

Fig. 86

(10.1) Definition. Let M^n be a connected n-dimensional manifold and $f, g: \mathbb{R}^n \to M^n$ two embeddings. We say that f and g are *compatibly oriented* if either M^n is not orientable or, f and g, relative to fixed orientations of $\mathbb{R}^n$ and M^n, are both either orientation preserving or reversing.

(10.2) Remark. If $\tau: \mathbb{R}^n \to \mathbb{R}^n$ is given by $\tau(x_1, \ldots, x_n) = (-x_1, x_2, \ldots, x_n)$ and $f, g: \mathbb{R}^n \to M^n$ are not compatibly oriented, then f and $g \circ \tau$ are.

(10.3) Lemma. *If two embeddings of $\mathbb{R}^n$ in the connected n-dimensional manifold M^n are compatibly oriented, then they are isotopic.*

Proof. Let f and g be the two embeddings. First, we want to convince ourselves that, w.l.o.g., we may take $f(0) = g(0)$.

On a connected manifold for any two points p and q there always exists a diffeotopy H, which takes p into $q: H_1(p) = q$. One only needs to embed an isotopy between the embeddings

$$\{p\} \to \{p\} \subset M$$

and

$$\{p\} \to \{q\} \subset M$$

in some diffeotopy by means of (9.5), and each differentiable path from p to q gives us such an isotopy, see Fig. 87. If now H is a diffeotopy with

98

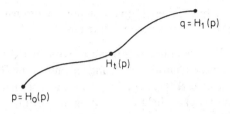

Fig. 87

$H_1(f(0)) = g(0)$, then it is enough to show that $H_1 \circ f$ and g are isotopic, since isotopy is an equivalence relation. Since it is clear that all $H_t \circ f$ are compatibly oriented, so are $H_1 \circ f$ and g, so that the problem is reduced to the case $f(0) = g(0)$. *We shall now therefore assume that $f(0) = g(0)$.*

The next step in the proof will be to 'shrink' f and g. But before doing that, we wish to make a short remark about $\mathbb{R}^n$, which will also be frequently useful later. Given prescribed $r_0 > 0$, $\epsilon > 0$ we choose a C^∞-function ϕ on $[0, \infty)$ with everywhere positive slope, which is given by $\phi(r) = r$ on $[0, r_0]$ and whose limit as $r \to \infty$ is $r_0 + \epsilon$, see Fig. 88. Then, if $\psi(r) = (1/r) \phi(r)$, ψ is also a C^∞-function on $[0, \infty)$, see Fig. 89, and

$$\sigma_t(x) := \psi(t|x|) \cdot x$$

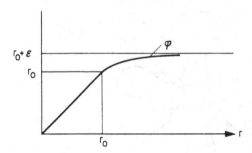

Fig. 88

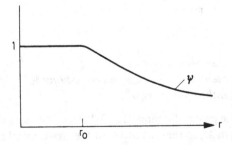

Fig. 89

defines an isotopy σ of embeddings $\mathbb{R}^n \to \mathbb{R}^n$ (work in polar coordinates!), of which we wish to collect some properties for future use.

(10.4) Assertion. For prescribed $r_0 > 0$ and $\epsilon > 0$ there exists an isotopy σ (shrinking) between the identity on $\mathbb{R}^n$ and an embedding $\mathbb{R}^n \to \mathbb{R}^n$, with image $(r_0 + \epsilon)\overset{\circ}{D}{}^n = \{x \in \mathbb{R}^n | |x| < r_0 + \epsilon\}$, which is such that all points of $r_0 D^n = \{x \in \mathbb{R}^n | |x| \leqslant r_0\}$ are held fixed during the isotopy, see Fig. 90.

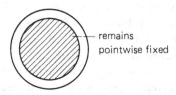

remains pointwise fixed

Fig. 90

In particular, σ_1 is a diffeomorphism between $\mathbb{R}^n$ and $(r_0 + \epsilon)\overset{\circ}{D}{}^n$, which is the identity on $r_0 D^n$. For example, we have:

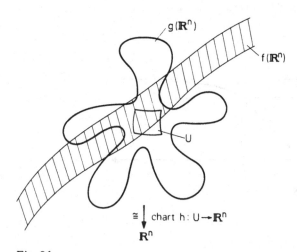

$g(\mathbf{R}^n)$

$f(\mathbf{R}^n)$

U

$\cong$ chart $h: U \to \mathbf{R}^n$

$\mathbf{R}^n$

Fig. 91

(10.5) Corollary. *If an open neighbourhood of $r_0 D^n \subset \mathbb{R}^n$ is embedded in a manifold M, then there also exists an embedding $\mathbb{R}^n \to M$, which agrees with the given embedding on $r_0 D^n$.*

Now we continue with the proof of lemma (10.3). We can choose a chart about the point $f(0) = g(0)$ in M, so that the image of the chart domain U is all of $\mathbb{R}^n$ see Fig. 91. This is easy to do, since $\epsilon \overset{\circ}{D}{}^n \cong \mathbb{R}^n$.

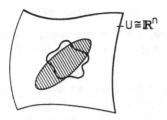

$-U \cong \mathbb{R}^n$

Fig. 92

Next, we can choose a shrinking (10.4) sufficient to ensure that $f \circ \sigma_1(\mathbb{R}^n)$ $\subset U$ and $g \circ \sigma_1(\mathbb{R}^n) \subset U$, see Fig. 92. Since $f \circ \sigma_1$ is isotopic to f (the isotopy is given by $h_t := f \circ \sigma_t$), and likewise $g \circ \sigma_1$ to g, we have yet to show that $f \circ \sigma_1$ and $g \circ \sigma_1$ are isotopic.

It is now enough to consider embeddings $\mathbb{R}^n \to \mathbb{R}^n$, and we pick one such, $\phi : \mathbb{R}^n \to \mathbb{R}^n$, with $\phi(0) = 0$. Then (and this is the essential point of the whole proof) ϕ is isotopic to the linear embedding $D\phi_0 : \mathbb{R}^n \to \mathbb{R}^n$, given by the Jacobi matrix at the point zero.

In fact lemma (2.3) there exist differentiable maps $\psi_i : \mathbb{R}^n \to \mathbb{R}^n$, $i = 1, 2, \dots, n$, with $\phi(x) = \sum_{i=1}^{n} x_i \psi_i(x)$, and then the Jacobi matrix consists precisely of the columns $\psi_i(0)$:

$$D\phi_0 = (\psi_1(0), \dots, \psi_n(0)).$$

One now defines the isotopy between ϕ and $D\phi_0$ by

$$(t, x) \mapsto \sum_{i=1}^{n} x_i \psi_i(tx) = \begin{cases} \phi(tx)/t & \text{for } t > 0 \\ D\phi_0 \cdot x & \text{for } t = 0. \end{cases}$$

$\underbrace{\qquad\qquad}_{\substack{\text{clearly} \\ \text{differentiable}}}$ $\underbrace{\qquad\qquad\qquad}_{\substack{\text{clearly an} \\ \text{embedding } \mathbb{R}^n \to \mathbb{R}^n \text{ for each } t.}}$

If now two linear embeddings (hence isomorphisms) $\mathbb{R}^n \to \mathbb{R}^n$ are compatibly oriented, then they are in the same connected component of $GL(n, \mathbb{R})$, and are therefore isotopic (the elementary transformations of a matrix — adding a multiple of a row (column) to another, multiplying a row (column) by some number $\alpha \neq 0$ — do not change the path component if $\alpha > 0$).

In the case of an oriented manifold M we can now complete the proof of lemma (10.3): here not only f and g but also $f \circ \sigma_1$ and $g \circ \sigma_1$ are compatibly oriented, both with respect to M and also with respect to $U \cong \mathbb{R}^n$. Therefore we obtain isotopic Jacobi matrices, since they have the same orientation.

If, however, M is not orientable, so that f and g cannot be assumed to satisfy an orientation condition, then $f \circ \sigma_1$ and $g \circ \sigma_1$ may be oppositely oriented with respect to $U \cong \mathbb{R}^n$. At first, therefore, the route via Jacobi matrices seems blocked.

Clearly, this problem is solved if we can prove the following:

Proposition. *If M is a connected, non-orientable manifold and $p \in M$ then there exists a diffeotopy H of M with $H_1(p) = p$, such that $T_p H_1: T_p M \to T_p M$ is orientation reversing.*

Let us suppose that this proposition is false. Then we could choose an orientation for $T_p M$ and orient every other tangent space $T_q M$ in the following way: choose a differentiable path $\alpha: [0, 1] \to M$, $\alpha(0) = p$, $\alpha(1) = q$; embed it in a diffeotopy H^α and orient $T_q M$ via

$$T_p H_1^\alpha: T_p M \cong T_q M.$$

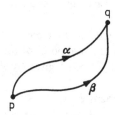

Fig. 93

This orientation of $T_q M$ is indeed independent of the choice of α and H^α because, if another diffeotopy H^β induced the opposite orientation, then the composition of H^α with the reversed diffeotopy associated with H^β (see (9.2)) would have the property required by the proposition, see Fig. 93. So in this way we would obtain an orientation of M, which was assumed non-orientable — contradiction.

In this way the proposition, and with it the lemma (10.3), is proved.   □

(10.6) Definition. Let M_1 and M_2 be n-dimensional connected manifolds — oriented in the orientable case. Let

$$f_1: \mathbb{R}^n \to M_1$$
$$f_2: \mathbb{R}^n \to M_2$$

be embeddings — if the manifolds are oriented assume f_1 preserves the orientation and f_2 reverses it. Then one calls the n-dimensional manifold, which is obtained from the disjoint union

$$[M_1 - f_1(D^n/3)] + [M_2 - f_2(D^n/3)]$$

by the identification of $f_1(tx)$ with $f_2((1 - t)x)$ for all $1/3 < t < 2/3$, $x \in S^{n-1}$, the *connected sum* of M_1 and M_2 relative to the embeddings f_1 and f_2, denoted by $M_1 \# M_2$, see Fig. 94.

Before we make ourselves more familiar with the connected sum, this is perhaps the place to make some general remarks about 'identification'.

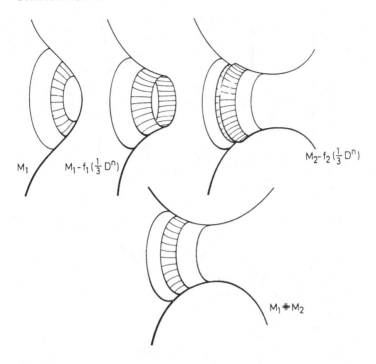

Fig. 94

(10.7) Remarks about identification. Let X and Y be topological spaces, $X_0 \subset X$, $Y_0 \subset Y$ subspaces and $\alpha: X_0 \to Y_0$ a homeomorphism. Then one can glue X and Y by means of α along X_0 and Y_0 to obtain a new topological space $X \cup_\alpha Y$. Thus:

In $X + Y$, one introduces an equivalence relation $\sim$ by setting each point $x_0 \in X_0$ equivalent to its image point $\alpha(x) \in Y_0$. The equivalence classes take the form

$$\{x\} \text{ for } x \in X - X_0,$$

$$\{y\} \text{ for } y \in Y - Y_0,$$

and

$$\{x, \alpha(x)\} \text{ for } x \in X_0.$$

The set $X + Y/\sim$ of equivalence classes, equipped with the quotient topology, is then denoted by $X \cup_\alpha Y$, see Fig. 95.

Assertion. One can canonically consider X and Y as subspaces of $X \cup_\alpha Y$.

Assertion. If X and Y are differentiable manifolds, X_0 and Y_0 open submanifolds, $\alpha: X_0 \to Y_0$ a diffeomorphism and (!) $X \cup_\alpha Y$ a Hausdorff space, then $X \cup_\alpha Y$ is again in canonical fashion a differentiable manifold.

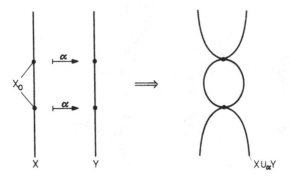

Fig. 95

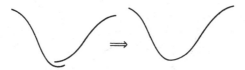

Fig. 96

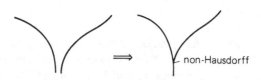

Fig. 97

So, for example, if instead of the identification $f_1(tx) \mapsto f_2((1-t)x)$, illustrated in Fig. 96, we take the identification $f_1(tx) \mapsto f_2(tx)$, illustrated in Fig. 97, we do not obtain a manifold (even though the identification space is still locally Euclidean).

The condition that $X \cup_\alpha Y$ is again Hausdorff can be stated thus: if $x \in X - X_0$, $x_\nu \in X_0$ and $\lim (x_\nu) = x$, then $\lim (\alpha(x_\nu))$ does not exist in Y_0.

(10.8) Assertion and orientation convention. A connected sum of connected manifolds M_1 and M_2 is orientable precisely when M_1 and M_2 are orientable, and there then exists exactly one orientation on $M_1 \# M_2$ which is compatible with the given orientations on $M_i - f_i(D^n/3)$, $i = 1, 2$. From now on, a connected sum of oriented manifolds will always be given this orientation.

The construction of $M_1 \# M_2$ uses embeddings $f_i \colon \mathbb{R}^n \to M_i$. That such embeddings always exist (assuming that M_i is non-empty) is obvious (charts and (10.5)). To what extent however is $M_1 \# M_2$ independent of the choice of these embeddings?

First of all, the following is clear: if $f_i: \mathbb{R}^n \to M_i$ and $f_i': \mathbb{R}^n \to M_i'$ are such embeddings and $\phi_i: M_i \stackrel{\approx}{\to} M_i'$ diffeomorphisms, for which

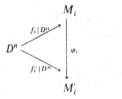

is commutative, then the ϕ_i induce a diffeomorphism between $M_1 \# M_2$ (formed using f_1 and f_2) and $M_1' \# M_2'$ (formed using f_1' and f_2').

In the case $M_i = M_i'$, we know already that f_i and f_i' are isotopic because of the assumed compatibility of orientation (lemma (10.3)). This isotopy, however, does not necessarily fix all points outside a compact subset of $\mathbb{R}^n$, and so, without further discussion, we cannot embed it in a diffeotopy. We would like to do this, because then

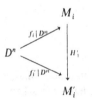

would be commutative. It is, however, also unnecessary to embed the whole isotopy in a diffeotopy, since we only use it on D^n.

(10.9) Complement to the isotopy theorem. *If h is a technical isotopy of embeddings $M \to N$ and $M_0 \subset M$ is compact, then there is a diffeotopy H of N, which is fixed outside a compact subset of N, with $h_t | M_0 = H_t \circ h_0 | M_0$.*

Proof. The proof proceeds almost exactly as that of the isotopy theorem itself (9.5), except that the required vector field X on $\mathbb{R} \times N$ has to satisfy condition (ii'):

$$T_{(t,x)} F(\partial/\partial t) = X(F(t,x))$$

only for points $(t,x) \in \mathbb{R} \times M_0$. The argument on p. 95, in which the independence of h_t from t outside M_0 plays a role, is now dispensable. $\square$

(10.10) Corollary. *The (where relevant, oriented) diffeomorphism type of $M_1 \# M_2$ does not depend on the choice of embeddings $\mathbb{R}^n \to M_i$.*

We can therefore, in cases where it is only a matter of (where relevant, oriented) diffeomorphism types, simply speak of 'the' connected sum $M_1 \# M_2$, while having in mind some particular connected sum.

(10.11) Exercises

1 Let M be an oriented connected manifold, $p, q \in M$ and $\phi\colon T_p M \cong T_q M$ an orientation preserving isomorphism. Show that there exists a diffeomorphism $f\colon M \to M$ with $T_p f = \phi$.

2 Show that there exists no embedding $f\colon \mathbb{R}^2 \to S^1 \times \mathbb{R}$, for which $f(\mathbb{R}^2)$ contains one of the sets $S^1 \times \{x\}$, see Fig. 98.

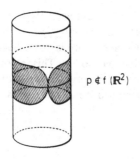
$p \notin f(\mathbb{R}^2)$

Fig. 98

Hint: use exercise 13 in Chapter 9.

3 That two arbitrary embeddings of $\mathbb{R}^n$ in an n-dimensional, non-orientable, connected manifold are isotopic has a remarkable consequence: in the case that our universe, which we only know locally, is globally not diffeomorphic to $\mathbb{R}^3$, but, for example, to $S^1 \times \mathbb{R}\mathbf{P}^2$, then one would be able to make a journey from which one's mirror image would return (heart on the right-hand side, etc.). Try to believe it!

4 If one picks a base point from each of k copies of S^n, and passes from the disjoint union $S^n + \ldots + S^n$ to the quotient space obtained by identifying these k points, one obtains a so called 'bouquet' of k n-spheres. Describe a subspace of $\mathbb{R}^{n+1}$, which is homeomorphic to this bouquet of spheres. Is the bouquet a manifold?

5 Let $\mathfrak{A} = \{h_\alpha\colon U_\alpha \to U'_\alpha \,|\, \alpha \in A\}$ be an atlas for a topological n-dimensional manifold M. Consider the finest equivalence relation on the topological disjoint sum $\Sigma_{\alpha \in A} U'_\alpha$, under which two points are equivalent if they correspond to each other under some change of chart. Show that $\Sigma_{\alpha \in A} U'_\alpha / \sim$ is homeomorphic to M.

6 Show that

$$(M_1 \# M_2) \# M_3 \cong M_1 \# (M_2 \# M_3),$$

$$M_1 \# M_2 \cong M_2 \# M_1,$$

and

$$M \# S^n \cong M,$$

where the various connected sums may be defined using (10.6) and (10.8).

7 $\mathbb{R}^n \# \ldots \# \mathbb{R}^n \cong ? \subset \mathbb{R}^n$.

8 Show that $\mathbb{R}\mathbf{P}^2 \# \mathbb{R}\mathbf{P}^2$ admits a nowhere vanishing vector field.

9 Show that if M_1 and M_2 are compact submanifolds of $\mathbb{R}^k$, then $M_1 \# M_2$ is also embeddable in $\mathbb{R}^k$.

10 If n is odd, then $\mathbb{R}\mathbf{P}^n$ is orientable. Show that the diffeomorphism type of $\mathbb{R}\mathbf{P}^n \# M$ is independent of which orientation one chooses in the two summands.

11 Let $M_1, \ldots, M_k$ be connected n-dimensional manifolds. Show that

is diffeomorphic to $M_1 \# M_2 \# \ldots \# M_k \# (S^1 \times S^{n-1})$, see Fig. 99.

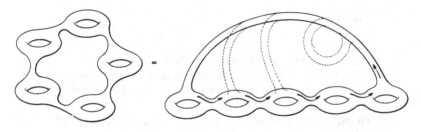

Fig. 99

11

Second order differential equations and sprays

If M is an open subset of $\mathbb{R}^n$, the straight line from each point $x \in M$, $t \mapsto x + tv$, with prescribed velocity, remains with M for some time (Fig. 100) and any two points in M which are sufficiently close to each other can be joined by such a path.

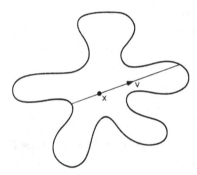

Fig. 100

On a general manifold, one can of course do the same thing locally with the help of charts, but for global problems this is worthless, since the connecting paths of course depend on the charts, and so in the regions of overlap are not well defined.

For example, if M is open in $\mathbb{R}^n$ and $f, g : X \to M$ are close in the C^0-topology, then a homotopy between f and g in M is defined by

$$(x, t) \mapsto (1 - t)f(x) + tg(x), \text{ see Fig. 101.}$$

In order to imitate such a construction for a general manifold M, we need a coordinate free substitute for the connecting paths between two points. This is the concern of the present chapter.

Traditionally this is carried out 'quite simply': one introduces a Riemannian metric on M; locally the geodesics play the role of straight lines. For a book such as this one, this has the disadvantage that one must assume a knowledge

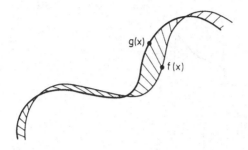

g(x)

f(x)

Fig. 101

of Riemannian geometry. Therefore, we follow instead the method of *sprays*, applied by S. Lang in [3], and which can be completely developed in a few pages.

(11.1) Notation. We recall, once more, that for a differentiable curve $\gamma: (a, b) \to M$ in a manifold, we denote by $\dot{\gamma}(t) \in T_{\gamma(t)}M$ the *velocity vector* of the curve:

$$\dot{\gamma}(t) := T_t\gamma(d/dt).$$

The *velocity curve* $\dot{\gamma}: (a, b) \to TM$ is then a differentiable curve in TM, for which we can again apply the same notation:

$$\ddot{\gamma}: (a, b) \to TTM$$

is the velocity curve of $\dot{\gamma}$, where TTM denotes the tangent bundle of the total space of the tangent bundle of M.

(11.2) Definition. A *second order differential equation* on a manifold M is a vector field ξ on TM with the property that every solution curve β of ξ is the velocity curve of its projection on M, that is, $\beta = \dot{\gamma}$ for $\gamma = \pi \cdot \beta$, see Fig. 102.

(11.3) Definition. A curve $\gamma: (a, b) \to M$ is called a *solution curve* of the second order differential equation ξ on M if $\dot{\gamma}$ is the solution curve of ξ on TM, that is, if for all t

$$\ddot{\gamma}(t) := \xi(\dot{\gamma}(t))$$

Since the solution curves of ξ on TM and M are related by the formulae

$$\gamma = \pi \cdot \beta, \quad \beta = \dot{\gamma},$$

we can consider them as two ways of looking at one and the same thing.

The definition of a second order equation on M as a vector field on TM corresponds to the more familiar use of the same notation in calculus. Thus, the second order equation

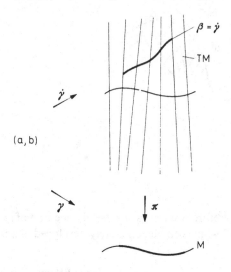

Fig. 102

$$y'' = f(y, y')$$

is equivalent to the first order system

$$y' = z, \quad z' = f(y, z).$$

(11.4) Notation. If ξ is a second order differential equation on M, then for each $v \in TM$ the associated maximal solution curve of ξ in TM will be denoted by β_v, and the projection $\pi \cdot \beta_v$ on M by γ_v.

Thus, for $v \in T_x M$ the curve $\gamma_v : (a_v, b_v) \to M$ is the maximal solution curve of ξ in M with $\gamma_v(0) = x$ and $\dot{\gamma}_v(0) = v$, see Fig. 103.

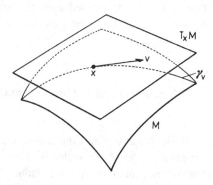

Fig. 103

Such a curve $t \mapsto \gamma_v(t)$ will be the substitute for the straight line $t \mapsto x + tv$ in $\mathbb{R}^n$. But, in order to make such a substitute geometrically usable, one will have to demand, at least, that γ_v and γ_{sv} differ from each other only in the velocity of passage (in contrast to say ballistics, where different solution curves are associated with different initial velocities along the same direction, see Fig. 104).

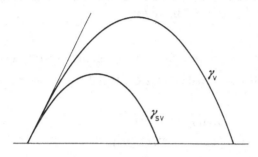

Fig. 104

(11.5) Definition. A second order differential equation ξ on M is called a *spray* if for $s, t \in \mathbb{R}$, $v \in TM$, the number t belongs to the domain of definition of γ_{sv} if and only if st belongs to the domain of definition of γ_v, and if in this case

$$\gamma_{sv}(t) = \gamma_v(st).$$

(11.6) Theorem on the existence of sprays. *On every manifold there exists a spray.*

Proof. Until now we have stated the conditions on a vector field ξ on TM to be a second order differential equation and a spray as conditions on the solution curves. What do they say directly about ξ?

Assertion 1. *A vector field ξ on TM is a second order differential equation if and only if $T\pi \circ \xi = \mathrm{Id}_{TM}$:*

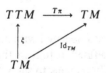

For if ξ is a second order differential equation, $v \in TM$, β_v is the solution curve for v in TM and $\gamma_v := \pi \circ \beta_v$ the solution curve in M, then

$$T\pi \cdot \xi(v) = T\pi(\dot{\beta}_v(0)) = \dot{\gamma}_v(0) = \beta_v(0) = v,$$

or $T\pi \cdot \xi = \mathrm{Id}_{TM}$. If, the other way round, ξ is the vector field with $T\pi \cdot \xi = \mathrm{Id}_{TM}$, then for the flow lines β we have

$$\beta(t) = T\pi \cdot \xi(\beta(t)) = T\pi(\dot{\beta}(t)) = \dot{\gamma}(t).$$

This checks the second order condition on β, and completes the proof of the first assertion.

Assertion 2. *A second order differential equation ξ on M is a spray if and only if for all $s \in \mathbb{R}$, and $v \in TM$, we have*

$$\xi(sv) = Ts(s\xi(v)),$$

where $Ts\colon TTM \to TTM$ denotes the differential of multiplication by s.

F   if ξ is a spray, then for fixed $s \in \mathbb{R}$, $v \in TM$ and for t allowed to vary in a neighbourhood of zero,

$$\gamma_{sv}(t) = \gamma_v(st) \Rightarrow \dot{\gamma}_{sv}(t) = s\dot{\gamma}_v(st) \Rightarrow$$
$$\beta_{sv}(t) = s\beta_v(st) \Rightarrow \dot{\beta}_{sv}(t) = Ts(s\dot{\beta}_v(st)).$$

Hence for $t = 0$:

$$\xi(sv) = Ts(s\xi(v)),$$

which is the required condition.

Conversely, let ξ be a second order differential equation which satisfies this equation, and let

$$\gamma_v\colon (a_v, b_v) \to M$$

be the maximal solution curve in M with initial velocity v. We show first that the equation $\alpha(t) := \gamma_v(st)$ gives a solution curve with initial velocity sv. To this end we check that

$$\dot{\alpha}(0) = s\dot{\gamma}_v(0) = s\beta_v(0) = sv,$$

the correct initial value and, moreover,

$$\dot{\alpha}(t) = s\dot{\gamma}_v(st).$$

Therefore, $\ddot{\alpha}(t) = Ts(s\ddot{\gamma}_v(st)) = Ts(s\xi(\dot{\gamma}_v(st)))$ which, by the assumed formula, however, equals

$$\xi(s\dot{\gamma}_v(st)) = \xi(\dot{\alpha}(t)).$$

Therefore, $\ddot{\alpha}(t) = \xi(\dot{\alpha}(t))$, and α is the solution curve associated with initial velocity sv. For all values of t, for which $\gamma_v(st)$ is defined, it follows that $\gamma_{sv}(t)$ is also defined and that $\gamma_v(st) = \gamma_{sv}(t)$. It only remains to show that if $\gamma_{sv}(t)$ is defined, then so is $\gamma_v(st)$. For $s \neq 0$ we have only to apply the argument above with $1/s$ instead of s; for $s = 0$ it is clear in any case, because each solution curve is defined at the point zero. This concludes the proof of the second assertion.

We now turn to the construction of a spray on a given manifold M. The conditions from assertions 1 and 2, thus

$$T\pi \cdot \xi = \mathrm{Id}_{TM} \quad \text{and} \quad \xi(sv) = Ts(s\xi(v)) \quad \text{for all } s, v,$$

which ξ must satisfy, are conditions on the restrictions $\xi \,|\, T_xM$, which must be satisfied for each $x \in M$, see Fig. 105. Both are clearly 'convex' conditions in the sense that, given two sections ξ_1 and ξ_2 of $TTM \,|\, T_xM$, which satisfy the conditions, then so does $(1 - \lambda)\xi_1 + \lambda\xi_2$. Hence it is enough to show that each point in M has a neighbourhood U, on which there exists a spray, for we can then glue such local sprays together to form a global spray on M by means of a partition of unity.

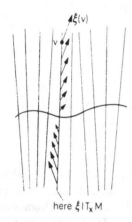

here $\xi \,|\, T_x M$

Fig. 105

For the local problem, we are justified in taking U as an open subset of $\mathbb{R}^n$. We can therefore write

$$TU = U \times \mathbb{R}^n, \quad TTU = U \times \mathbb{R}^n \times \mathbb{R}^n \times \mathbb{R}^n,$$

which is to so arrange things that the velocity curve of a curve

$$t \mapsto (x(t), v(t)) \in TU = U \times \mathbb{R}^n$$

is given by

$$t \mapsto (x(t), v(t), dx/dt(t), dv/dt(t)).$$

Since $\pi \colon TU \to U$ is given by $(x, v) \mapsto x$, and hence $T\pi \colon TTU \to TU$ by $(x, v, w, b) \mapsto (x, w)$, one writes the differential of multiplication by s

$$Ts \colon TTU \to TTU \quad \text{as} \quad (x, v, w, b) \mapsto (x, sv, w, sb).$$

A second order differential equation is therefore a section $\xi \colon TU \to TTU$ of the form $\xi(x, v) = (x, v, v, \psi(x, v))$. Translated into the usual terminology of the infinitesimal calculus, the differential equation becomes $y'' = \psi(y, y')$, and such a differential equation is a spray precisely if $\psi(x, sv) = s^2 \psi(x, v)$.

If, as in our case, no further conditions need to be imposed on the spray, then we have, for example, in

$$\xi: TU \to TTU \quad (x, v) \mapsto (x, v, v, 0),$$

found a spray on U, see Fig. 106. With this therefore we have also proved theorem (11.6).   □

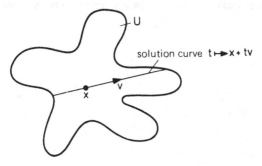

solution curve $t \mapsto x + tv$

Fig. 106

(11.7) Exercises

1 Let (E, π, M) be a differentiable vector bundle. If one restricts $T\pi: TE \to TM$ to $TE \,|M\, (M = \text{zero-section!})$, then one has a bundle homomorphism $TE\,|M \to TM$. Show that this bundle homomorphism is surjective, and that the subbundle $E \subset TE\,|M$ is its kernel.

2 Let (E, π, M) be a differentiable vector bundle. Prove that $TE \cong \pi^*E \oplus \pi^*TM$.

3 Give an example of a non-trivial differentiable vector bundle E, whose tangent bundle TE is trivial.

4 Let M be a non-empty connected manifold. Show that there exists a differentiable curve $\gamma: \mathbb{R} \to M$, so that the image of the velocity curve $\dot\gamma: \mathbb{R} \to TM$ is dense in TM.

5 Construct a spray for $M = S^1$, for which not all maximal solution curves are defined on all of $\mathbb{R}$.

6 Let M be a manifold, $\dim M \geqslant 1$. Show that not every curve in M can arise as the solution curve of a second order differential equation.

7 Give an example of a spray on S^n (as a vector field on $TS^n \subset S^n \times \mathbb{R}^{n+1}$), whose solution curves are great circles.

12
The exponential map and tubular neighbourhoods

(12.1) Remark. Let ξ be a spray on M. Then the set

$$\mathscr{O}_\xi := \{v \in TM \mid \gamma_v(1) \text{ is defined}\}$$

is an open neighbourhood of the zero-section in TM.

Proof. We denote the maximal flow on TM, whose velocity field is ξ, by Φ and its domain of definition by $A \subset \mathbb{R} \times TM$. Therefore

$$\mathscr{O}_\xi = \{v \in TM \mid (1, v) \in A\},$$

is open (compare (8.11)) because A is open. Moreover, from $\xi(sv) = Ts(s\xi(v))$, putting $s = 0$ one sees that ξ vanishes on the zero-section, and thus the flow lines of the points of the zero-section (as fixed points) are defined for all $\mathbb{R}$, in particular for $t = 1$. Therefore Φ_ξ contains the zero-section. $\square$

(12.2) Definition. If ξ is a spray on M, then the map

$$\exp_\xi: \mathscr{O}_\xi \to M, v \mapsto \gamma_v(1)$$

is called the *exponential map* of ξ, see Fig. 107.

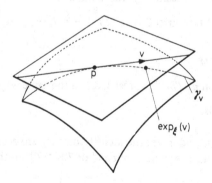

Fig. 107

Clearly, $\exp_\xi$ is a differentiable mapping because, if Φ is the flow associated to ξ, then $\exp_\xi$ is given by $v \mapsto \pi \cdot \Phi(1, v)$. We now want to determine the differential

115

$$T_p \exp_\xi : T_p TM \to T_p M$$

of $\exp_\xi$ at the points of the zero-section $M \subset TM$. (Since $\mathcal{O}_\xi$ is open in TM, $T_p \mathcal{O}_\xi = T_p TM$.)

To this end, first let us agree on a notation. If E is a differentiable vector bundle over M and $p \in M$ is a point of the zero-section, then $T_p E$ has two significant subspaces $T_p E_p$ and $T_p M$; for both E_p and M (= zero-section) are submanifolds of E (Fig. 108), which pass through p. By looking at a bundle chart we see that $T_p E$ is actually the direct sum of $T_p E_p$ and $T_p M$ and, since $T_p E_p$ is canonically isomorphic to E_p, we have $T_p E = E_p \oplus T_p M$ for each $p \in M$. Globally we have $TE | M = E \oplus TM$.

(12.3) Notation. If E is a differentiable vector bundle over M, then in the canonical isomorphism

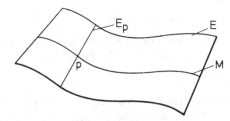

Fig. 108

$$TE | M = E \oplus TM$$

we wish to keep to this order of summands, so that even in the case

$$TTM | M = TM \oplus TM$$

there will be no confusion as to the meaning of the summands.

(12.4) Remark. The differential $T\exp_\xi : TTM \to TM$, restricted to $TTM | M = TM \oplus TM$ is

$$(\text{Id}, \text{Id}) : TM \oplus TM \to TM.$$

The differential of the projection $\pi : TM \to M$, restricted in the same way, is

$$(0, \text{Id}) : TM \oplus TM \to TM.$$

Proof. Both maps, $\exp_\xi$ and π, are the identity on the zero-section M, from which it follows that on the second summand of $TM \oplus TM$ both their differentials are the identity.

Now let v be a vector from the first summand; then v is the velocity vector of the curve $t \mapsto tv$ in TM, $\mathcal{O}_\xi$ respectively, at the point $t = 0$, see Fig. 109. The image curve under the projection is constant, therefore $T\pi(v) = 0$. However, the image curve under the exponential map is $t \mapsto \exp_\xi(tv) = \gamma_{tv}(1) = \gamma_v(t)$, and therefore $T \exp_\xi(v) = \dot\gamma_v(0) = v$. $\quad\square$

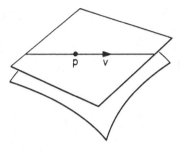

Fig. 109

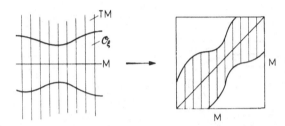

Fig. 110

(12.5) Corollary. *On the zero-section the differential of the map* $(\pi, exp_\xi)\colon \mathcal{O}_\xi \to M \times M$ *is given by*

$$
\begin{array}{|c|c|}
\hline
0 & Id \\
\hline
Id & Id \\
\hline
\end{array}
\colon T_p M \oplus T_p M \to T_p M \oplus T_p M = T_{(p,p)} M \times M
$$

In particular, on the zero-section the map has maximal rank.

From now on in this chapter we shall be much concerned with maps of this kind. An important geometric consequence of the property of having maximal rank on the zero-section is formulated in the following lemma:

(12.6) Lemma. *Let M be an n-dimensional manifold, (E, π, X) a differentiable vector bundle with n-dimensional total space E. Let U be an open neighbourhood of the zero-section in E (see Fig. 111) and f: U → M a differentiable map which has maximal rank on the zero-section and which also embeds the zero-section X in M. Then there is an open neighbourhood U_0 of the zero-section in U, so that $f \mid U_0$ is an embedding and, therefore, here it is a diffeomorphism onto an open neighbourhood of f(X) in M (see Fig. 112).*

Proof. We may assume that f is everywhere of maximal rank on U

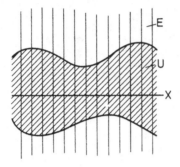

Fig. 111

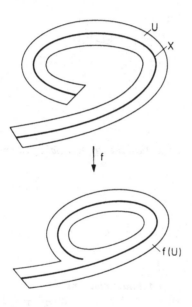

Fig. 112

(5.3), then $f: U \to f(U)$ is open and a local homeomorphism. We may further assume that $f(U) = M$, and the embedding $f \mid X$ allows us to consider X as a subset of M. We are therefore looking for a local inverse to the map $f: U \to M$ near X. For the proof we recall the following lemma from general topology (which is familiar in sheaf theory, see Godement [1], p. 150):

(12.7) **Section extension lemma.** *Let* $f: U \to M$ *be a local homeomorphism,* $X \subset M$ *a subset such that each neighbourhood of* X *in* M *contains a paracompact neighbourhood (this holds in particular for manifolds and, more generally, for metric spaces* M*). Let* $s: X \to U$ *be a section of* f*, that is,*

$f \circ s = \mathrm{Id}_X$. *Then there exists an open neighbourhood W of X in M, and an extension of s to a section s: W → U, and s(W) =: U_0 is open in U.*

Proof of (12.7). In M we may choose a family $\{V_\lambda\}_{\lambda \in \Lambda}$ of open sets, which cover X, and sections $s_\lambda: V_\lambda \to U$ of f, which are such that $s_\lambda | V_\lambda \cap X = s | V_\lambda \cap X$. This is possible because f is a local homeomorphism. Now we may assume that the V_λ cover all of M, that M is paracompact (replace M by some neighbourhood of X), that this covering is locally finite, and that it admits a refinement $\{W_\lambda\}_{\lambda \in \Lambda}$ with $\bar{W}_\lambda \subset V_\lambda$ (Kelley [8], chapter 5, theorem 28, p. 156; for manifolds see (7.1)).
Now we put

$$W := \{x \in M \mid x \in \bar{W}_\lambda \cap \bar{W}_\mu \Rightarrow s_\lambda(x) = s_\mu(x)\}.$$

Then, clearly, $X \subset W$, and we have extended the section $s: X \to U$ continuously to W. It therefore only remains to show that W (respectively $s(W)$) is a neighbourhood of X.

Suppose then that $x \in X$. We choose a neighbourhood Q of $s(x)$, which is mapped homeomorphically by f onto a neighbourhood of x, see Fig. 113. Next we choose a neighbourhood A of x in M, which is so small, that

   (i) $A \subset f(Q)$,
   (ii) A intersects only finitely many $\bar{W}_\lambda$, say $\bar{W}_1, \ldots, \bar{W}_k$,
   (iii) $x \in \bar{W}_i$, $i = 1, \ldots, k$,
   (iv) $A \subset V_i$, $i = 1, \ldots, k$,
   (v) $s_i(A) \subset Q$, $i = 1, \ldots, k$.

Then $s_1 | A = \ldots = s_k | A = (f|Q)^{-1} | A$, and so from (ii) $A \subset W$.

With this we have proved (12.7) and hence (12.6).   □

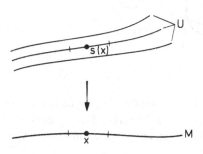

Fig. 113

In this proof we have followed S. Lang [3]. In the literature one frequently finds other proofs of (12.6), which use a somewhat complicated topological argument, but which do not generalise to infinite dimensional

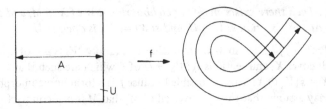

Fig. 114

manifolds. Moreover, at this point, the following assertion easily slips into the argument (we have found it in four books): if $f: U \to M$ is a local homeomorphism, $A \subset U$ is closed, and $f \mid A: A \to f(A)$ is injective, then it is possible to extend f to a homeomorphism of a neighbourhood of A.

Counterexample (Fig. 114): $U = (0, 1) \times (0, 1); M = \mathbb{R}^2, A = (0, 1) \times \frac{1}{2}$. The mistake lies in the assumption that, given the hypotheses, $f \mid A: A \to f(A)$ is a local homeomorphism.

Next, in order to make the application of (12.6) easier, we wish to remark that a 'nice' neighbourhood of the zero-section is contained in every preassigned neighbourhood of the zero-section, see Fig. 115.

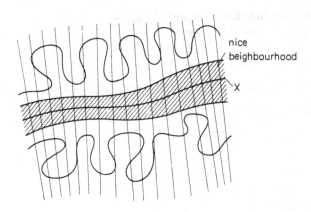

Fig. 115

(12.8) Remark. If (E, π, X) is a differentiable vector bundle with a Riemannian metric $\langle \, , \rangle$ and if U is a neighbourhood of the zero-section, then there exists a differentiable everywhere positive function $\epsilon: X \to \mathbb{R}$, so that the open neighbourhood

$$\epsilon \mathring{D} E = \{v \in E \mid |v| < \epsilon(\pi(v))\}$$

is contained in U.

Proof. Locally this is clearly possible, even with constant ϵ (see Fig.

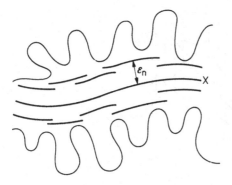

Fig. 116

116): For an appropriate cover, one chooses a subordinate partition of unity $\{\tau_n \mid n \in \mathbb{N}\}$ and obtains a global ϵ in the form $\epsilon = \Sigma_{n \in \mathbb{N}} \epsilon_n \cdot \tau_n$. $\qquad\square$

As a first application of the exponential map and of the lemma (12.6) we prove

(12.9) Theorem. *Let M be a manifold and Y a topological space. If two continuous maps*

$$f, g \colon Y \to M$$

are sufficiently close in the C^0-topology (compare (7.8)), then they are homotopic, that is, there exists a continuous map $h \colon [0, 1] \times Y \to M$ with $h(0, y) = f(y)$ and $h(1, y) = g(y)$ for all $y \in Y$.

Proof. We choose a spray on M and a Riemannian metric for TM. Then for the exponential map

$$\exp \colon \mathcal{O} \to M$$

of the spray, we choose a small positive function ϵ, so that $\epsilon \mathring{D} TM \subset \mathcal{O}$, and so that

$$(\pi, \exp) \mid \epsilon \mathring{D} TM$$

is a diffeomorphism onto an open neighbourhood U of the diagonal $\mathbf{\Delta}_M$ in $M \times M$. All this is possible by (4.20), (11.6), (12.5), (12.6), (12.8).

Observe that the diagram

$$
\begin{array}{ccc}
\epsilon \mathring{D} TM & \xrightarrow{\ (\pi, \exp)\ } & U \subset M \times M \ni (p, q) \\
{\scriptstyle \pi}\big\downarrow & \big\downarrow & \big\uparrow \\
M & \xrightarrow{\hspace{2cm}} & \Delta_M \qquad (p, p)
\end{array}
$$

is commutative; therefore the points $(\pi, \exp)^{-1}(p, q)$ all lie in the fibre over p.

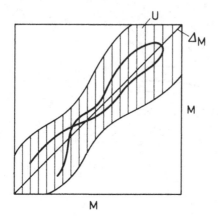

Fig. 117

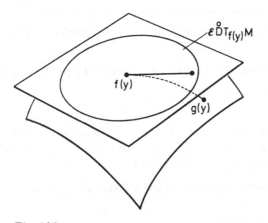

Fig. 118

Next, if $f, g: Y \to M$ are sufficiently close in the C^0-topology (compare (7.8)), then $(f(y), g(y))$ must lie in U for all $y \in Y$ (that is, close to $(f(y), f(y)) \in \Delta_M$), see Fig. 117.

If we set

$$h(t, y) := \exp\left(t((\pi, \exp)\,|\,\epsilon \overset{\circ}{D}TM)^{-1}(f(y), g(y)))\right)$$

we have found the required homotopy, see Fig. 118.   □

Now we wish to turn to tubular neighbourhoods. In the study of sub-manifolds $X \subset M$ we must often handle problems which, while not local and so not transferable by means of a chart to a problem in $\mathbb{R}^n$, do not involve the whole manifold M, but only the consideration of a neighbourhood of the submanifold. For such considerations it is therefore very useful to know that the 'position' of X in such a neighbourhood is the 'same' as the 'position' of

X as zero-section in its normal bundle. The following definition makes this precise:

(12.10) Definition. If $X \subset M$ is a submanifold, then by a *tubular map* for X one understands an embedding

$$\tau: \bot X \to M$$

of the normal bundle $\bot X$ of X into M, which on X is the inclusion $X \subset M$, and for which the differential induces the identity $\bot X \to \bot X$ on the zero-section.

The differential of τ, restricted to $(T \bot X) \mid X$, is a bundle homomorphism

$$\bot X \oplus TX \to TM \mid X$$

(compare (12.3)), because τ restricted to X is the inclusion. The condition stated last in the definition concerns itself with the composition

$$\bot X \oplus 0 \xrightarrow{T\tau} TM \mid X \xrightarrow{\text{Proj.}} \bot X = (TM \mid X)/TX.$$

(12.11) Theorem on the existence of tubular maps. *For every submanifold there exists a tubular map.*

Proof. Let $X \subset M$ be a submanifold. We choose a spray on M with an exponential map exp: $\mathcal{O} \to M$, and choose a Riemannian metric $\langle \, , \, \rangle$ for TM. By means of the canonical isomorphism $\bot X = (TX)^{\bot}$ we consider $\bot X$ as a subbundle of $TM \mid X$. Then on the neighbourhood

$$U := \mathcal{O} \cap \bot X$$

of the zero-section in $\bot X$, a map

$$U \to M$$

is given by the exponential map, which is the inclusion $X \subset M$ on X. Since the differential of the exponential map, restricted to $TTM \mid M$ is exactly (Id, Id): $TM \oplus TM \to TM$ (12.4), then the differential of exp $\mid U: U \to M$, restricted to $(T \bot X) \mid X = \bot X \oplus TX$ is just the identity

$$\bot X \oplus TX \xrightarrow{\cong} TM \mid X.$$

From this we draw two conclusions: first, the differential has maximal rank on the zero-section and thus fulfills the hypotheses of (12.6) and, second, it induces the identity $\bot X \to \bot X$.

Next, we choose a small positive function ϵ on X, so that $\epsilon \mathring{D} \bot X \subset U$ and so that exp $\mid \epsilon \mathring{D} \bot X$ is an embedding ((12.6), (12.8)).

Finally, we choose a diffeomorphism

$$\bot X \xrightarrow{\cong} \epsilon \mathring{D} \bot X,$$

which on $(\epsilon/2) \cdot \mathring{D} \bot X$ is the identity (compare the technique applied for (10.4)). Then, clearly, the composition

$$\bot X \to \epsilon \mathring{D} \bot X \xrightarrow{\text{exp}} M$$

is a tubular map. □

(12.12) **Definition.** If $\tau: \bot X \to M$ is a tubular map for $X \subset M$, and if $\bot X$ is equipped with a Riemannian metric $\langle , \rangle$, then the neighbourhood

$$\tau(D \bot X)$$

of X in M is called a *tubular neighbourhood* of X.

Therefore the tubular map τ equips the tubular neighbourhood with the structure of a *'disc bundle'* so that with the given tubular map one also speaks of *fibres* and of the *projection*

tubular neighbourhood $\to X$ (see Fig. 119).

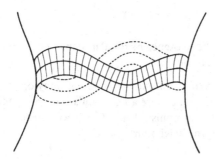

Fig. 119

For certain constructions which make use of all of this structure, it is important to know how far the construction depends on the choice of the tubular neighbourhood. To this end one has the following uniqueness theorem with which we want to end this section:

(12.13) **Uniqueness theorem for tubular neighbourhoods of compact submanifolds.** *Let X be a compact submanifold of a manifold M; $\tau_0, \tau_1: \bot X \to M$ tubular maps; $\langle , \rangle_0$ and $\langle , \rangle_1$ Riemannian metrics on $\bot X$; and finally let $U_0 := \tau_0 (D_0 \bot X)$ and $U_1 := \tau_1(D_1 \bot X)$ be the associated tubular neighbourhoods of X. Then there exists a diffeotopy H of M, which is fixed on X and which is such that H_1 maps the tubular neighbourhood U_0 fibrewise onto U_1. Furthermore, it is even possible to choose H, so that all points outside a compact subset of M are likewise held fixed, and so that for each point $p \in X$ and each t, $T_p H_t$ induces the identity $\bot_p X \to \bot_p X$.*

Proof. Clearly, it is enough to prove the theorem for the two special cases

(a) $\tau_0 = \tau_1 =: \tau$

and

(b) $\langle\, ,\rangle_0 = \langle\, ,\rangle_1 := \langle\, ,\rangle$.

For (a): we only need to find a fibre preserving isotopy h of the identity on $\perp X$, which leaves a neighbourhood of the zero-section fixed, and for which h_1 maps the disc bundle $D_0 \perp X = \{v \in \perp X \mid v\mid_0 \leqslant 1\}$ onto $D_1 \perp X$. Then we can embed the isotopy of $\tau \mid D_0 \perp X$, given by

$$\tau \circ h_t$$

in a diffeotopy H with the required properties (see 10.9).

W.l.o.g., we may suppose that $|v\mid_0 \leqslant |v\mid_1$ for all $v \in \perp X$. If $\phi: \mathbb{R} \to [0, 1]$ is a C^∞-function of the kind illustrated in Fig. 120 (compare Chapter 7), then an isotopy of the required kind is given by

$$h(t, v) := \left[\phi(t\mid v\mid_0)\frac{|v\mid_1}{|v\mid_0} + (1 - \phi(t\mid v\mid_0))\right] v$$

(Naturally one sets $h(t, 0) = 0$.) This proves case (a).

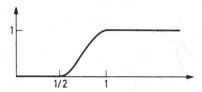

Fig. 120

For (b): here it is enough to find an isotopy between τ_0 and τ_1, so that each τ_t is a tubular map. We can forget the metric on $\perp X$. Instead of this we choose a metric $\langle\, ,\rangle$ for TM, a spray on M and $\epsilon > 0$, so that

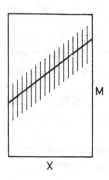

Fig. 121

$$(\pi, \exp): \epsilon \mathring{D}(TM \mid X) \to X \times M$$

is an embedding, see Fig. 121.

Next we must make use of the possibility of shrinking a differentiable vector bundle: there always exists a fibre preserving isotopy of the identity, which leaves fixed a neighbourhood of the zero-section, and whose end-embedding maps the whole bundle into a preassigned neighbourhood of the zero-section.

Seizing this possibility for $\perp X$ we recognise that we may already take τ_0 and τ_1 to be so small, that $(\pi, \tau_0)(\perp X)$ and $(\pi, \tau_1)(\perp X) \subset X \times M$ are contained in $(\pi, \exp)(\frac{1}{2}\epsilon \mathring{D}(TM \mid X))$. Therefore, both

and
$$\tau_0' := (\pi, \exp)^{-1}(\pi, \tau_0)$$
$$\tau_1' := (\pi, \exp)^{-1}(\pi, \tau_1)$$

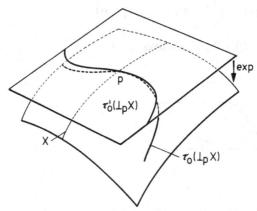

Fig. 122

are fibre preserving maps

$$\perp X \to TM \mid X$$

which, after composition with the exponential map, give τ_0 and τ_1, see Fig. 122.

It is now enough to show that τ_0' and τ_1' are connected by a fibre preserving isotopy which keeps the zero-section fixed, which for each p and t induces the identity on $\perp_p X$ via $T\tau_p'$, and which takes place completely in $\epsilon \mathring{D}TM \mid X$. Then $\tau := \exp \cdot \tau'$ will do what is required.

We can forget the condition that τ' takes place in $\epsilon \mathring{D}TM \mid X$: if the argument works anywhere in $TM \mid X$, it will work equally well in $\epsilon \mathring{D}TM \mid X$ (shrinking argument).

Since we now have all of $TM \mid X$ to play with, we can replace τ_0' and τ_1' by the bundle maps given by their differentials at the zero-section

$$\tau_0'': \perp X \to TM \mid X$$

$$\tau_1'': \perp X \to TM \mid X.$$

This 'linearisation' proceeds exactly as in the proof of (10.3) by means of lemma (2.3):

$$[0, 1] \times \perp X \to TM \mid X$$

$$(t, v) \mapsto \begin{cases} \dfrac{\tau_0'(tv)}{t} & \text{for } t \neq 0 \\ \\ T_p \tau_0'(v) & \text{for } t = 0 \end{cases}$$

is the isotopy between τ_0' and τ_0'', which we need. The argument for τ_1' is analogous.

To finish the argument, compose both τ_0'' and τ_1'' with the projection

$$\perp X \xrightarrow{\tau_i'} TM \mid X \xrightarrow{\text{Proj.}} (TM \mid X)/TX = \perp X, \; i = 0, 1.$$

This gives the identity of $\perp X$, something which holds also for each

$$(1 - t)\tau_0'' + t\tau_1''.$$

With

$$\tau'': [0, 1] \times \perp X \to TM \mid X$$

$$(t, v) \mapsto (1 - t)\tau_0''(v) + t\tau_1''(v)$$

we obtain an isotopy with the required properties. $\qquad\qquad \square$

(12.14) Exercises

1 Let $X \subset M$ be a submanifold whose normal bundle possesses a section everywhere different from zero. Show that the inclusion $X \subset M$ is isotopic to an embedding whose image is disjoint from X.

2 Show that two disjoint closed submanifolds of M also have disjoint tubular neighbourhoods.

3 Let X be a submanifold of M. Show that if X is compact and $M - X$ connected, then so is the complement of every tubular neighbourhood of X in M. The hypothesis that X is compact is not superfluous.

4 Let M be a connected manifold and $X \subset M$ a (codimension one) connected submanifold. If X lies 'one sidedly' in M, that is, the normal bundle $\perp X$ is not trivial, show that $M - X$ is connected.

5 Let M be a manifold. Show that a connected subset $X \subset M$ is a submanifold, provided that there exists an open neighbourhood U of X and a differentiable map $f: U \to U$ with $f \circ f = f$ and $f(U) = X$.

6 Let X be a closed k-codimensional submanifold in M with a trivial normal bundle. Show that there is a differentiable map

$$f: M \to S^k,$$

so that X is the pre-image of a regular value of f.

7 Let (E, π, M) be a differentiable vector bundle. Then the set $P(E)$ of 1-dimensional subspaces of the fibres is in a canonical manner a

manifold and, over $P(E)$, we have a canonical differentiable line bundle

$$\eta(E) \to P(E),$$

for which the fibre over a point $p = V \subset E_x$ of $P(E)$ is the line $\{p\} \times V$. Clearly, we have a canonical linear map of vector bundles $\eta(E) \to E$ (one says that $\eta(E)$ arises from E by 'blowing up' the zero-section). Show that there is a diffeomorphism

$$\eta(E) - 0\text{-section} \xrightarrow{\cong} E - 0\text{-section}$$

given by the canonical map $\eta(E) \to E$.

8 Let $X \subset M$ be a compact submanifold and τ a tubular map for X. Show that there exists exactly one differentiable structure on $(M - X) \cup P(\perp X) =: M_X$, for which the maps (i) and (ii) below are embeddings:

(i) $M - X \subset M_X$

(ii) $\eta(\perp X) \to M_X$

$$v \mapsto \begin{cases} v & \text{for } v \in P(\perp X) = \text{zero-section of } \eta(\perp X) \\ \tau(\phi(v)) & \text{for } v \in \eta(\perp X) - \{\text{zero-section}\}, \end{cases}$$

where $\phi: \eta(\perp X) \to \perp X$ is the canonical map. Show that the differentiable structure of M_X does not depend on the choice of tubular map. (One says that the differentiable manifold M_X arises from M by 'blowing up' X.)

9 Show that the blowing up of a codimension one submanifold has no effect.

10 Show that by blowing up a point in S^n, one obtains the projective space $\mathbb{R}\mathbf{P}^n$. (In general, blowing up a point of M^n is, up to diffeomorphism, the same as taking the connected sum $M \# \mathbb{R}\mathbf{P}^n$.)

11 Construct a non-empty, n-dimensional manifold $M, n \geq 2$, for which the blowing up of a point does not change the diffeomorphism type.

13
Manifolds with boundary

Manifolds, which are locally modelled on Euclidean space, are not the only interesting geometric objects one can imagine. However, without further assumptions one cannot base the theory developed so far on local models other than Euclidean space, even if a corresponding generalisation of manifold were easy to define. The basic methods which we have learnt rest, namely, on the possibility of performing analysis on manifolds (differential equations, inverse functions, etc.), and here the essential local statements depend on properties of Euclidean space.

However, one can extend many methods from the theory of manifolds to spaces, which are built up from local models other than Euclidean space, as long as these spaces or local models are sufficiently sensibly composed or built up from manifolds ('stratified spaces'). We shall not go into this, restricting ourselves to the classical and simplest case of manifolds with boundary, which locally look like the closed Euclidean half-space

$$\mathbb{R}^n_+ := \{x \in \mathbb{R}^n \mid x_n \geqslant 0\} \quad \text{(Fig. 123)}.$$

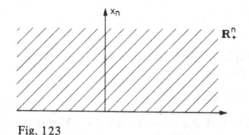

Fig. 123

These manifolds with boundary are important, not only as a generalisation but, also, as an aid in the theory of 'ordinary' manifolds.

Since it makes sense (as is well known from infinitesimal calculus) to speak of C^∞-maps defined on open subsets of $\mathbb{R}^n_+$, there is no difficulty in replacing $\mathbb{R}^n$ everywhere by $\mathbb{R}^n_+$ in the definition of a differentiable manifold. Since we want to call on this analogy several times from now on, we shall explicitly write down the definition on this first occasion:

129

(13.1) Definition. A *topological n-dimensional manifold with boundary* is a second countable Hausdorff space M, which is locally homeomorphic to $\mathbb{R}^n_+$. An atlas of local charts

$$h: U \to U'$$

(U open in M, U' open in $\mathbb{R}^n_+$, h homeomorphism) is called *differentiable* if the chart transformations are differentiable; and an *n-dimensional differentiable manifold with boundary* is a pair consisting of a topological n-dimensional bounded manifold M and a maximal differentiable atlas $\mathfrak{D}$ for M.

The rank theorem easily gives an example:

(13.2) Note. If M is an (ordinary) manifold and $a \in \mathbb{R}$ a regular value of $f: M \to \mathbb{R}$, then $f^{-1}((-\infty, a])$ is canonically a manifold with boundary, see Fig. 124. About a point $p \in f^{-1}(a)$, one can clearly choose $a-f$ as last coordinate of a chart.

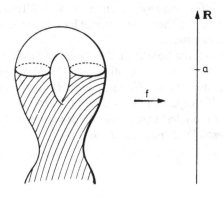

Fig. 124

In this way many examples of manifolds with boundary present themselves to us, for example the disc

$$D^n = \{x \in \mathbb{R}^n \mid |x|^2 \leqslant 1\}$$

or, more generally, for a differentiable vector bundle (E, π, X) with Riemannian metric $<, >$ and a positive differentiable function ϵ on X, the *ϵ-disc bundle ϵDE*

$$\epsilon DE := \{v \in E \mid |v|^2 \leqslant \epsilon^2(\pi(v))\}.$$

A diffeomorphism of one open subset of $\mathbb{R}^n_+$ onto another, maps each point of the 'boundary' (that means, each point with $x_n = 0$) to a point on the boundary, because an invertible germ $(\mathbb{R}^n, x) \to (\mathbb{R}^n, y)$ possesses an open representative, and hence cannot take an 'interior' point to a point on

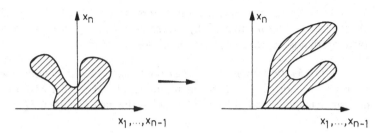

Fig. 125

the boundary, see Fig. 125. It follows that the boundary of a manifold with boundary is well defined and can itself be given the structure of a differentiable manifold.

(13.3) Definition and Notation. If M is an n-dimensional manifold with boundary, then a point $p \in M$, which is mapped by some (and hence by every) chart about p to a point with $x_n = 0$, is called a *boundary point* of M. The set of boundary points of M is canonically an $(n-1)$-dimensional manifold (in the usual sense) which we shall denote by ∂M, and call the *boundary* of M (Fig. 126). $M - \partial M$ is canonically an (ordinary) n-dimensional manifold, and is called the *interior* of M.

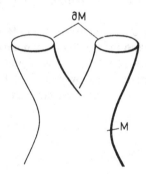

Fig. 126

(13.4) Convention. In order to avoid always having to speak of 'ordinary' instead of the newly introduced manifolds with boundary, we wish to agree that manifolds with boundary will always be called manifolds with boundary, and that the word 'manifold' will be reserved for the usual, unbounded manifolds. However, it will be possible for a manifold with boundary to have an empty boundary. If M is a manifold with boundary and $\partial M = \emptyset$, then $M = M - \partial M$ is of course also canonically a manifold. By a *closed* manifold we understand a compact manifold (without boundary).

A manifold with boundary is formed from the two manifolds $M - \partial M$ and ∂M. We have, therefore, first of all, to describe how these two manifolds fit together, that is, to describe a neighbourhood of ∂M in M.

(13.5) Definition. By a *collar* for a manifold with boundary we mean a diffeomorphism from the manifold with boundary $\partial M \times [0, 1)$ onto an open neighbourhood of ∂M in M, which is the inclusion $\partial M \subset M$ on ∂M, see Fig. 127.

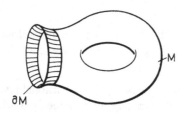

Fig. 127

(13.6) Theorem. *Every manifold with boundary has a collar.*

Proof. Note that one can consider the boundary as a submanifold and therefore obtain the collar as half a tubular neighbourhood. In detail, and in a simpler fashion, we argue as follows:

For manifolds with boundary, one defines the tangent bundle (TM, π, M) as for unbounded ones, and indeed so that also for the boundary points, $T_x M$ is a vector space, rather than just a half-space. (To apply the 'geometer's definition' meaningfully here would be rather clumsy; however, the definitions of the 'algebraist' or the 'physicist' carry over word for word, compare (2.2) or

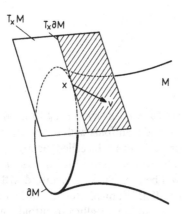

Fig. 128

(2.5).) For $x \in \partial M$, $T_x \partial M$ is a 1-codimensional subspace of $T_x M$, which decomposes $T_x M$ into two half-spaces, of which, relative to some and hence to every chart about x, one lies on the side of the manifold. We wish to call a vector $v \in T_x M$, which is not tangential to ∂M and which belongs to this half-space, an *inward pointing* vector, see Fig. 128.

Pointing inward is a convex property in the sense already used several times. Therefore, by means of a partition of unity, we can easily construct a vector field X on M, so that each vector of $X \mid \partial M$ points inward. Then there exists a positive function ϵ on M, and a differentiable map of

$$\{(x, t) \in \partial M \times \mathbb{R}_+ \mid 0 \leqslant t < \epsilon(x)\}$$

to M, which for each fixed x is a solution curve of X with initial value x. This map is the inclusion on ∂M, it is injective, it is of maximal rank everywhere and, therefore, as can be easily seen, it is a diffeomorphism onto an open neighbourhood of ∂M in M, see Fig. 129. By means of 'shrinking'

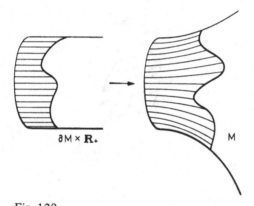

$\partial M \times \mathbb{R}_+$ M

Fig. 129

(compare (10.4)) we easily obtain a diffeomorphism of $\partial M \times [0, 1)$ — indeed, if we so wish, we can map $\partial M \times \mathbb{R}_+$ onto a neighbourhood of ∂M in M, which is the inclusion $\partial M \subset M$ on M. $\qquad \square$

For collars, as for tubes, there exists a uniqueness theorem which, here (for the sake of simplicity) we only formulate and prove for compact boundaries.

(13.7) Theorem. *If M is a manifold with compact boundary, and κ_0, κ_1 are two collars for M, and K is a compact neighbourhood of ∂M in M, then there exists $\epsilon > 0$ and a diffeotopy of M, which leaves ∂M and the complement of K pointwise fixed, and which on $\partial M \times [0, \epsilon)$ takes the collar κ_0 into κ_1.*

Proof. We construct a family X_λ of vector fields, depending differentiably on λ, on a neighbourhood of ∂M in M as follows: The vector field

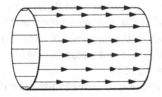

Fig. 130

$\partial/\partial t$ on $\partial M \times [0, 1)$ (Fig. 130) is taken by κ_0 and κ_1 into two vector fields, both defined in a neighbourhood of ∂M in M. Label these fields X_0 and X_1 on the intersection U of the two neighbourhoods (Fig. 131). Then let us define $X_\lambda := (1 - \lambda)X_0 + \lambda X_1$ on U.

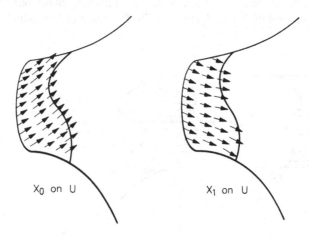

X_0 on U X_1 on U

Fig. 131

Along ∂M each X_λ points inward. By integration we therefore obtain, for sufficiently small ϵ, an isotopy κ between κ_0 and κ_1 on $\partial M \times [0, 2\epsilon)$. Here each κ_λ is a collar and, indeed, the whole isotopy takes place in the (topological) interior $\mathring{K}$ of K. As in the isotopy theorem (complement (10.9)) we now find a diffeotopy of $\mathring{K}$, which leaves fixed all points outside a compact subset of K, and in which $\kappa \,|[0, 1] \times \partial M \times [0, \epsilon]$ is embedded. We extend this diffeotopy to one of M by decreeing all points outside of $\mathring{K}$ to be held fixed. In this way we have found the required diffeotopy. $\square$

In order to prepare a first application of collars, just look at the following situation: let N be a manifold and $\tau : N \to N$ at a fixed point free involution, that is, a diffeomorphism with $\tau(p) \neq p$ for all p, and $\tau \cdot \tau = \mathrm{Id}_N$.

If one identifies points which correspond to each other under τ and denotes the quotient space by N/τ, then the canonical projection $N \xrightarrow{\;\pi\;} N/\tau$ is topologically a two leaved covering and, because τ is a diffeomorphism, there exists exactly one differentiable structure on N/τ with respect to which π is a local diffeomorphism. We therefore consider N/τ as a differentiable manifold. Example: $\tau\colon S^n \to S^n, x \mapsto -x$, then $S^n/\tau = \mathbb{R}\mathbf{P}^n$.

(13.8) **Definition and Notation.** Let M be a manifold with boundary, $\tau\colon \partial M \to \partial M$ a fixed point free involution, and κ a collar for M. Then there exists exactly one differentiable structure on the (unbounded) topological manifold M/τ, which is obtained by identifying points which correspond to each other under τ, with respect to which the canonical inclusion $M - \partial M \subset M/\tau$ and the map

$$\frac{\partial M \times (-1, 1)}{\tau \times (-\mathrm{Id})} \to \frac{M}{\tau}$$

$$[p, t] \mapsto \begin{cases} \kappa(p, t) & \text{for} \quad t \geqslant 0 \\ \kappa(\tau p, -t) & \text{for} \quad t \leqslant 0 \end{cases}$$

defined by κ are embeddings. The differentiable manifold defined in this way will also be denoted by M/τ.

The definition shows how one can use the canonical differentiable glueing of $\partial M \times [0, 1)$ to itself, giving $\partial M \times (-1, 1)/\tau \times (-\mathrm{Id})$ (locally this is illustrated in Fig. 132), by means of a collar to explain the differentiable identification space M/τ.

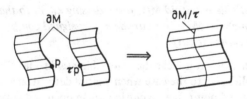

∂M   ∂M/τ

Fig. 132

The differentiable structure of M/τ indeed depends on the choice of the collar, as one recognises, for example, by considering the paths

$$(-1, 1) \to M/\tau$$

$$t \mapsto \begin{cases} \kappa(p, t) & t \geqslant 0 \\ \kappa(\tau p, -t) & t \leqslant 0, \end{cases}$$

which must be differentiable for each $p \in \partial M$. For example, if $M = \mathbb{R}_+^2 + \mathbb{R}_+^2$, hence $\partial M = \mathbb{R} + \mathbb{R}$, and τ is the natural interchange of the two boundary components, then M/τ is both as set and as topological manifold the same as $\mathbb{R}^2$. If one uses the natural collar $(x, t) \mapsto (x, t)$, then one obtains the differentiable structure of $\mathbb{R}^2$. If however, one chooses the collar given by

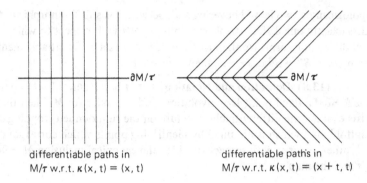

differentiable paths in
M/τ w.r.t. κ(x, t) = (x, t)

differentiable paths in
M/τ w.r.t. κ(x, t) = (x + t, t)

Fig. 133

$(x, t) \mapsto (x + t, t)$, then the paths $t \mapsto (x + |t|, t)$ become differentiable in $M/\tau = \mathbb{R}^2$, see Fig. 133. In actual fact the diffeomorphism type of M/τ does not depend on the collar, that is, two manifolds M/τ formed from two distinct collars are nonetheless diffeomorphic.

In the case of compact boundaries there even follows:

(13.9) Corollary from the uniqueness theorem for collars. *Let M be a manifold with (compact) boundary, $\tau: \partial M \to \partial M$ a fixed point free involution and κ_0, κ_1 collars for M. Then there is a diffeomorphism*

$$M/\tau \to M/\tau,$$

which takes the differentiable manifold M/τ, formed using κ_0, onto that formed using κ_1, and which on $\partial M/\tau$ and outside a preassigned compact neighbourhood of $\partial M/\tau$ is the identity.

(13.10) Explanation. It is clear that everything said so far about the construction of M/τ applies also to the case when τ is not defined on all of ∂M, but when $\tau: X \to X$ is a fixed point free involution on an open and closed subset X of ∂M (equal to a union of boundary components, see Fig. 134).

As a convention, we want to agree that, in cases where only the diffeomorphism type is important, given M and τ, we shall speak of 'the' differentiable manifold M/τ without specifying the collar.

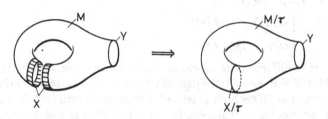

Fig. 134

As a special case of the construction, we consider the identification of two manifolds with boundary by means of a diffeomorphism of the boundaries.

(13.11) Definition. Let M_1, M_2 be manifolds with boundary, $X_i \subset \partial M_i$ be open and closed and $\phi: X_1 \xrightarrow{\cong} X_2$ be a diffeomorphism. Then we write

$$M_1 \underset{\phi}{\cup} M_2 := M/\tau,$$

where $M = M_1 + M_2$ and $\tau: X_1 + X_2 \to X_1 + X_2$ is given by $\tau \mid X_1 = \phi$, $\tau \mid X_2 = \phi^{-1}$, see Fig. 135.

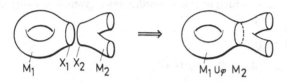

Fig. 135

The particular manifold without boundary $M \cup_{\mathrm{Id}} M$ which one obtains if one glues together two copies of M by means of Id: $\partial M \to \partial M$, is called the *double* of M.

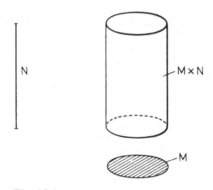

Fig. 136

As another application of the collar theorem, we shall show how one can present the product of two manifolds with boundary, again as a manifold with boundary. If M and N are manifolds with boundary, then $(M \times N) - (\partial M \times \partial N)$ has a canonical structure as a manifold with boundary, see Fig. 136. At the points of $\partial M \times \partial N$, from the charts for M and N, we obtain 'charts' for $M \times N$ which, instead of the half-space, map into open subsets of the 'quarter-space'

$$\mathbb{R}^m_+ \times \mathbb{R}^n_+ = \mathbb{R}^{m+n-2} \times \mathbb{R}_+ \times \mathbb{R}_+ \text{ (Fig. 137)}.$$

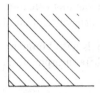

Fig. 137

In order to define a differentiable structure on all of $M \times N$, we use the homeomorphism of the half-plane $\mathbb{R}^2_+$ (see Fig. 138) onto the quadrant $\mathbb{R}_+ \times \mathbb{R}_+$ (see Fig. 139), which in polar coordinates is given by halving the angle. Denote this by ϕ:

$$\phi: (r, \theta) \mapsto (r, \theta/2).$$

ϕ defines a diffeomorphism $\mathbb{R}^2_+ - 0 \cong (\mathbb{R}_+ \times \mathbb{R}_+) - 0$.

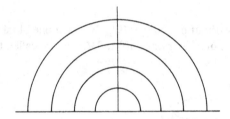

Fig. 138

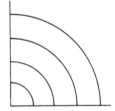

Fig. 139

(13.12) Definition and Notation. Let M and N be manifolds with boundary with collars κ and λ, taken here for technical reasons in the form

$$\kappa: \partial M \times \mathbb{R}_+ \to M$$

$$\lambda: \partial N \times \mathbb{R}_+ \to N.$$

Then there exists exactly one differentiable structure on $M \times N$, relative to which the maps

$$(M \times N) - (\partial M \times \partial N) \subset M \times N,$$

and

$$\partial M \times \partial N \times \mathbb{R}_+^2 \xrightarrow{\text{Id} \times \phi} \partial M \times \partial N \times \mathbb{R}_+ \times \mathbb{R}_+$$

$$\cong (\partial M \times \mathbb{R}_+) \times (\partial N \times \mathbb{R}_+) \xrightarrow{\kappa \times \lambda} M \times N$$

are embeddings, that is, are diffeomorphisms onto open subsets of $M \times N$. In future $M \times N$ is to be understood in this fashion as a differentiable manifold with boundary.

The technique used here is called 'straightening the angle'.

(13.3) Remark. The boundary of $M \times N$ is $\partial M \times N \cup_{\text{Id}_{\partial M \times \partial N}} M \times \partial N$ if one uses the collars given by κ and λ for $\partial M \times N$ and $M \times \partial N$, see Fig. 140. If one is only concerned with diffeomorphism type, then one can simply speak of 'the' product $M \times N$ as a differentiable manifold with boundary, without specifying the collars.

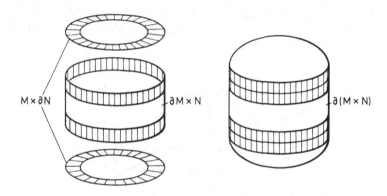

$M \times \partial N$ $\partial M \times N$ $\partial (M \times N)$

Fig. 140

We want to conclude this chapter on manifolds with boundary by introducing the notion of 'bordism', which plays so great a role in advanced differential topology.

Every manifold without boundary is the boundary of a manifold with boundary, for example $M = \partial(M \times [0, 1))$. But to be the boundary of a *compact* manifold with boundary is a restriction with interesting geometric consequences. More generally, one divides closed (that is, compact, un-bounded) manifolds into 'bordism classes' as follows:

(13.14) Definition. Two closed n-dimensional manifolds M_1 and M_2 are called *bordant* if there is a compact manifold with boundary W, such that $\partial W = M_1 + M_2$ (Fig. 141). If the closed manifold M is the boundary of a compact manifold with boundary, we call M *bounding* or *nullbordant*.

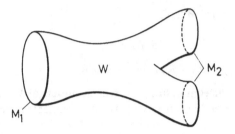

Fig. 141

(13.15) Remark and Definition. 'Bordant' is an equivalence relation. The equivalence classes are called *bordism classes*; we denote the bordism class of M by $[M]$.

Proof that 'bordism' is an equivalence relation. The clearly symmetric relation is, given $\partial(M \times [0, 1]) = M + M$, also reflexive. In order to understand transitivity, we apply the technique of glueing manifolds together: if $M_1 \sim M_2$ and $M_2 \sim M_3$, and if W_1, W_2 are compact manifolds with boundary with $\partial W_1 \cong M_1 + M_2$, $\partial W_2 \cong M_2 + M_3$, then $\partial(W_1 \cup_{\mathrm{Id}_{M_2}} W_2) = M_1 + M_3$. So transitivity follows, see Fig. 142.

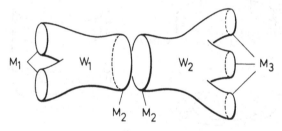

Fig. 142

(13.16) Remark. The disjoint sum of manifolds makes the set $\mathfrak{N}_n$ of bordism classes of n-dimensional manifolds into an abelian group; the Cartesian product defines a multiplication

$$\mathfrak{N}_n \times \mathfrak{N}_m \to \mathfrak{N}_{n+m},$$

with makes $\mathfrak{N}_* := \oplus_{n=0}^{\infty} \mathfrak{N}_n$ into a $\mathbb{Z}/2$-algebra.

One convinces oneself of this without difficulty or surprise. In contrast the structure of this algebra lies very deep.

The geometric-analytic techniques which we describe in this book, certainly form the basis of the study of differentiable manifolds, but they do not suffice to solve most of the harder problems. Here one also needs the help of algebraic topology. With the definition of the algebra $\mathfrak{N}_*$ we have come to a threshold between geometry and algebra. For quite a lot of geometric problems, which can be solved only with the help of algebraic topology, it is of great importance to know the structure of $\mathfrak{N}_*$. This structure was determined by R. Thom, who thereby laid the foundations for the extensive bordism theory. His result is:

(13.17) Theorem (Thom 1954). *Let* $\mathbb{Z}/2[X_2, X_4, X_5, X_6, X_8, X_9, \ldots]$ *be the polynomial ring over* $\mathbb{Z}/2$ *on countably many variables* X_i, *one for each* $i \geqslant 0$, *which is not of the form* $2^j - 1$. *Then there is an algebra isomorphism*

$$\mathbb{Z}/2[X_2, X_4, \ldots] \to \mathfrak{N}_*,$$

which maps each X_i *onto an element of* $\mathfrak{N}_i$. *One can so set up the isomorphism, that for each even* i, *the variable* X_i *is mapped onto the bordism class of the* i-*dimensional real projective space.*

(13.18) Exercises

1　Let M be a closed manifold and $a, b \in \mathbb{R}$ regular values of a differentiable function $f: M \to \mathbb{R}$. Show that the manifolds $f^{-1}(a)$ and $f^{-1}(b)$ are bordant.

2　Show that on each manifold with boundary M there exists a differentiable function with $f^{-1}(0) = \partial M$.

3　Show that $M - \partial M = \emptyset$ implies that $M = \emptyset$ also.

4　Show that an orientable manifold with boundary has a boundary which is also orientable.

5　Give an example of a manifold with (non-empty) boundary, whose diffeomorphism type is unaltered by the removal of an arbitrary point.

6　Let M be a compact manifold with boundary and X a vector field on M, which is inward pointing on the boundary. Show that $\mathbb{R}_+$ is contained in the domain of definition of each maximal solution curve.

7　Show that $M_1 \# M_2$ is bordant to $M_1 + M_2$.

8　Show that every closed manifold is bordant to a connected manifold.

9　Let A_0 and A_1 be disjoint closed subsets of the differentiable manifold M. Show that there exists a decomposition $M = M_0 \cup M_1$, $\partial M_0 = \partial M_1 = M_0 \cap M_1$ of M into two manifolds with boundary, which are glued along the common boundary, and which are such that $A_\nu \subset M_\nu$, see Fig. 143.

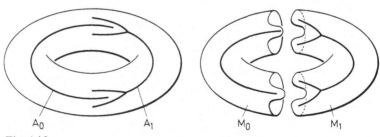

A_0　　　　　　A_1　　　　　M_0　　　　　M_1

Fig. 143

10 Show that the double of a compact manifold with boundary
    certainly bounds.

11 Let M be a compact manifold with boundary and $\phi: \partial M \to \partial M \times 0$
    the canonical diffeomorphism. Show that M is diffeomorphic to
    $M \cup_\phi (\partial M \times [0, 1])$, see Fig. 144.

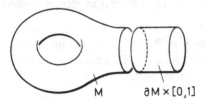

                               M $\partial M \times [0,1]$

Fig. 144

12 Show that for each diffeomorphism $\phi: S^{n-1} \to S^{n-1}$, the manifold
    $D^n \cup_\phi D^n$ is homeomorphic to S^n.

13 Give examples of oriented manifolds with boundary M_1 and M_2
    and of a diffeomorphism $\phi: \partial M_1 \overset{\cong}{\to} \partial M_2$, so that $M_1 \cup_\phi M_2$ is not
    orientable.

14 Show that a closed manifold, on which a fixed point free involution
    exists, necessarily bounds.

15 Show that $D^n \times D^m \cong D^{n+m}$.

16 How many bordism classes of 15-dimensional manifolds are there?
    Use (13.17).

14
Transversality

We study the following situation: let $f: M \to N$ be a differentiable map of differentiable manifolds, and let $L \subset N$ be a submanifold. What can we say about the pre-image $f^{-1}(L) \subset M$? If f is transverse to L, then we know that $f^{-1}(L) \subset M$ is a submanifold of the same codimension as L in N. However, without further hypotheses, $f^{-1}(L)$ has in general no structure of any kind, see Fig. 145.

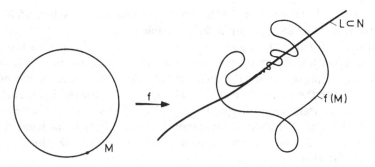

Fig. 145

(14.1) Theorem (Whitney). *Every closed subset of a differentiable manifold is the set of zeros of some differentiable function.*

Proof. Suppose, first, that A is a closed subset of the open subset V of $\mathbb{R}^m$, then we may cover the open set $V - A$ with a sequence of open discs $\{K_\nu \mid \nu \in \mathbb{N}\}$ and choose for each $\nu \in \mathbb{N}$ a differentiable function $\psi_\nu: V \to \mathbb{R}$ with the properties

(a) $0 \leqslant \psi_\nu$ and $\psi_\nu(x) > 0 \Leftrightarrow x \in K$,
(b) the absolute values of the functions ψ_ν and of all their derivatives, up to the νth order, are smaller than $2^{-\nu}$.

Condition (a) is easy to satisfy (Chapter 7); one satisfies condition (b) when one multiplies a function which satisfies (a), by a sufficiently small constant factor.

Now set $\psi := \Sigma_{\nu=1}^{\infty} \psi_{\nu}$. This sequence converges uniformly on all of V because of (b), as does the sequence for each derivative of the ψ_{ν}, and hence the limit function ψ is differentiable. Because of (a) $\psi(x) > 0$ if and only if $x \in K_{\nu}$ for some ν, that is, if and only if $x \notin A$.

More generally, suppose that $A \subset M$ is closed, and choose a partition of unity $\{\phi_i \mid i \in \mathbb{N}\}$, so that Supp (ϕ_i) is contained in a chart V_i for each i. Then Supp $(\phi_i) \cap A$ is closed in V_i and, as above, one finds a function $\lambda_i : V_i \to \mathbb{R}$, $\lambda_i \geqslant 0$, $\lambda_i(x) = 0 \Leftrightarrow x \in$ Supp $(\phi_i) \cap A$. Now set $\lambda = \Sigma_{i=1}^{\infty} \phi_i \lambda_i$ (with $\lambda_i = 0$ outside V_i).

The function λ is well defined and differentiable because the sum is locally finite. If $x \in A$, then $\lambda_i(x) = 0$ for all i, hence $\lambda(x) = 0$. If $x \notin A$, then $\phi_i(x) > 0$ for some i and $x \notin$ Supp $(\phi_i) \cap A$, hence $\lambda_i(x) > 0$, hence $\phi_i(x)\lambda_i(x) > 0$ and therefore $\lambda(x) > 0$. $\square$

(14.2) Remark. If we set $\alpha(x) := \exp(-\lambda(x)^{-2})$ with the function λ as constructed above, then $0 \leqslant \alpha < 1$, and $\alpha^{-1}(0) = A$. All derivatives of α vanish on A (with respect to any charts) because $\exp(-t^2)$ vanishes if and only if $t = 0$, and has a trivial Taylor Series at the zero point. Such functions are a useful technical aid.

Every closed subset $A \subset M$ is thus the pre-image of the submanifold $\{0\} \subset \mathbb{R}$ under an appropriate differentiable map. The situation is quite different for analytic or algebraic functions; for these there exists a large and interesting theory of zero sets for the appropriate functions. But the theory of pre-images of submanifolds under differentiable maps does not end here, since the peculiarly pathological maps, such as the one here constructed, are, in a certain sense, untypical exceptions – the usual state of a map is that of transversality. We shall show here – similarly to the immersion theorem – that one can approximate a map arbitrarily closely by a transverse map. First some preliminaries:

(14.3) Definition. We say that a vector bundle E is of *finite type* if E is a subbundle of a trivial bundle $\pi : B \times \mathbb{R}^k \to B$. In other words, there exists a vector bundle F over B so that $E \oplus F$ is trivial (4.11).

(14.4) Lemma. *A differentiable vector bundle over a differentiable manifold has finite type.*

Proof. The real reason is that the base is finite dimensional. In order not to have to load ourselves with too much topology, we permit ourselves the following argument: if a bundle has finite type, so clearly has every sub-bundle, likewise the restriction of the bundle to a subspace of the base. Moreover, the tangent bundle TM of a differentiable manifold has finite type, for the embedding $M \subset \mathbb{R}^n$ (following (7.10)) induces an inclusion $TM \subset T\mathbb{R}^n|M$, and the tangent bundle of $\mathbb{R}^n$ is trivial.

Now let $E \to N$ be some differentiable vector bundle, so that the total space E is a differentiable manifold and, as we have just said, the tangent bundle TE is of finite type. The same holds for the restriction of this bundle $TE | N \to N$ to the zero-section $N \subset E$. This bundle contains E as a subbundle, the normal bundle of the zero-section (12.3). □

The general transversality theorem depends on the following special case:

(14.5) Proposition. *Let (E, π, M) be a differentiable vector bundle which is equipped with a Riemannian metric. Let $N \subset E$ be a differentiable submanifold and ϵ a continuous everywhere positive function on M. Then there exists a differentiable section $s: M \to E$, $|s(p)| < \epsilon(p)$ for all $p \in M$, so that s is transverse to N. If $A \subset M$ is closed and the zero-section satisfies the transversality condition (5.11) with respect to N for all points of A, then one can choose the section s, so that $s | A = 0$ (Fig. 146).*

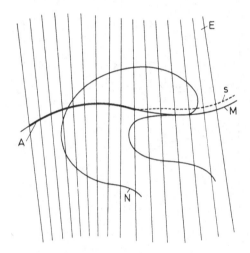

Fig. 146

Proof. Choose first a complement (E', π', M) to the vector bundle (E, π, M), so that $E \oplus E'$ is the trivial bundle $M \times \mathbb{R}^k$. Let $f: E \oplus E' \to E$ be the projection on the first factor, then the map $f: M \times \mathbb{R}^k \to E$ is a submersion, hence $f^{-1}(N) \subset M \times \mathbb{R}^k$ is a submanifold (for a submersion is certainly transverse), and the fibres of the normal bundle of $f^{-1}(N)$ in $M \times \mathbb{R}^k$ are mapped by Tf isomorphically onto the fibres of the normal bundle of N in E.

Hence a section s of the trivial bundle $M \times \mathbb{R}^k \to M$ is transverse to $f^{-1}(N)$ if and only if the section $f \circ s$ is transverse to N. To summarise: w.l.o.g., we may suppose that E is the trivial bundle $M \times \mathbb{R}^k \to M$. By the way, $f^{-1}(N)$ is the total space of the bundle $\pi^* E' | N$ over N.

Suppose, therefore, that $E = M \times \mathbb{R}^k$, that α is the function associated to the given closed set $A \subset M$ in (14.2), $U = M - A$ and $\delta = \epsilon \cdot \alpha: M \to \mathbb{R}$. Then $0 < \delta(p) < \epsilon(p)$ for all $p \in U$, and both δ and all its derivatives vanish on A. We have a bundle map

$$g: E \mid U \ = \ U \times \mathbb{R}^k \to U \times \mathbb{R}^k, \quad (p, v) \mapsto (p, (\delta(p))^{-1} v),$$

choose a regular value $w \in \mathbb{R}^k$, $|w| < 1$, of the composition

$$N \cap (E \mid U) \xrightarrow[g]{} U \times \mathbb{R}^k \xrightarrow[pr_2]{} \mathbb{R}^k,$$

and define the required section s by

$$s(p) \ = \ (p, \delta(p)w).$$

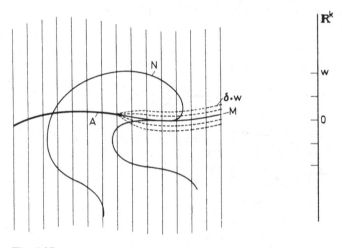

Fig. 147

We are here using Sard's theorem (6.1), see Fig. 147. At the point $p \in A$, the value of the function s and its differential agree with those of the zero-section; by hypothesis, transversality is satisfied. If $p \in U$, one has only to convince oneself that at p the section $g \cdot s \mid U$ (which has the constant value w) is transverse to $g(N \cap E \mid U)$, see Fig. 148. □

(14.6) Transversality theorem for sections (R. Thom). *Let $f: E \to M$ be a differentiable map between differentiable manifolds, and let $s: M \to E$ be a differentiable section of f (that is, $f \cdot s = \mathrm{Id}_M$). Let $N \subset E$ be a differentiable submanifold, then arbitrarily close to s there exists a section $t: M \to E$ transverse to N. If the transversality condition on s is already satisfied for all points of a closed subset $A \subset M$, then one can choose the section t so that $t \mid A = s \mid A$ (Fig. 149).*

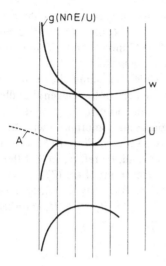

Fig. 148

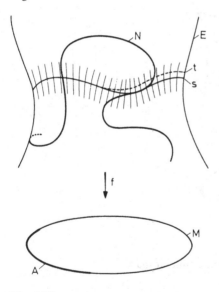

Fig. 149

'Arbitrarily close' is to be formulated with respect to some metric on E and the C^0-topology for maps, see (7.8).

Proof. We choose a well adapted tubular neighbourhood of $s(M)$, and apply proposition (14.5) in this tubular neighbourhood:

The section s is differentiable and an immersion, for $Tf \circ Ts = \mathrm{Id}$; also $s: M \to s(M)$ is a homeomorphism with inverse map $f \,|\, s(M)$ and so, following

(5.7), s is an embedding. Because $f|s(M)$ has rank equal to the dimension of M, f is a submersion in some neighbourhood U of $s(M)$, and it is enough to prove the theorem for $f: U \to M$, $s: M \to U$ and $N \cap U \subset U$. In other words, we may assume that f is a submersion ($U = E$).

Consider the bundle ker (Tf) over E which is a subbundle of the tangent bundle TE; then ker $(Tf)|s(M)$ is a complement of the tangent bundle of $s(M)$ in $TE|s(M)$ and therefore may serve as a normal component: The inclusion ker $(Tf)|s(M) \to TE|s(M)$ induces an isomorphism with the normal bundle of $s(M)$ in E. One can now define a spray $\xi: TE \to TTE$, so that $\xi(v) \in T(\text{ker }(Tf))$ for vectors $v \in$ ker (Tf) and, therefore, so that the integral curves, which begin in the direction of a vector out of ker (Tf), certainly preserve a direction from this subbundle. Put another way: the integral curves, which at one point are tangential to the 'fibre' $f^{-1}(p)$, $p \in M$, never leave $f^{-1}(p)$. The argument is illustrated by Figs. 149 and 150.

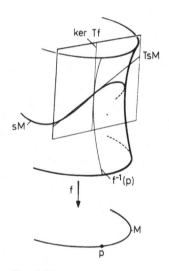

Fig. 150

From this spray one obtains a tubular map

$$\tau: \text{ker }(Tf)|s(M) \to E,$$

so that the diagram

$$
\begin{array}{ccc}
\text{ker}(Tf)|s(M) & \xrightarrow{\tau} & E \\
\pi \downarrow & & \downarrow f \\
s(M) & \xrightarrow[f]{\approx} & M
\end{array}
$$

commutes. Since τ is an open embedding, one can apply the proposition (14.5) directly to the left-hand side of the diagram. $\qquad\square$

As a special case we obtain the classical result of Thom:

(14.7) Transversality theorem for maps. *Let $f: M \to N$ be differentiable, and let $L \subset N$ be a differentiable submanifold. Then, arbitrarily close to f, there exist maps $g: M \to N$ transverse to L. If the transversality condition on f is already satisfied at the points of a closed subset $A \subset M$, then one can choose g so that $f | A = g | A$ (Fig. 151).*

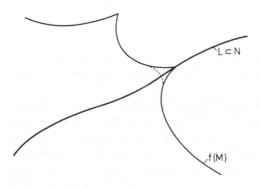

Fig. 151

Proof. Consider the composition $M \xrightarrow{s} M \times N \xrightarrow{\pi} N$, $s = (\mathrm{Id}, f)$, $\pi = $ projection, then $f = \pi \cdot s$, and π is a submersion, and hence transverse to L with the pre-image $\pi^{-1}(L) = M \times L \subset M \times N$. We may therefore approximate the section s of the projection $M \times N \to M$, following (14.6) by a section t transverse to $M \times L$. Hence – by the same conclusion as in the first step of the proposition – the map $\pi \cdot t: M \to N$ is transverse to $\pi(M \times L) = L$. $\qquad\square$

Note that, in this proof, we do not use the fact that the approximation t of s is a section. Without this condition (14.6) is much simpler to demonstrate because one can argue with a completely arbitrary tubular neighbourhood of $s(M)$.

Transversality theorems are basic to all 'general position' arguments in differential topology. With them begins cobordism theory, as well as the stability theory of differentiable maps, and they really explain why differential topology, far from being a desert of pathologies, yields a cornucopia of geometric phenomena.

As a topologist, one tries to approximate mappings between manifolds by others with good properties (differentiable, transverse, ...) because sufficiently close maps are homotopic (12.9). Suppose, therefore, $f_0, f_1: M \to N$ are sufficiently close approximations to a map f and are transverse to a submanifold

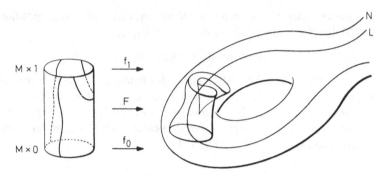

Fig. 152

$L \subset N$, then both are homotopic to f via differentiable homotopies
$M \times [0, 1] \to N$, $i = 0, 1$, which do not depend on t, wherever $f_0(p) = f(p)$
or $f_1(p) = f(p)$.

Let us choose a 'technical' homotopy which, for example, between the
times $0 \leqslant t \leqslant 1/3$ and $2/3 \leqslant t \leqslant 1$ is independent of t, then we can compose
the two original homotopies and obtain a differentiable (technical) homotopy
F between f_0 and f_1, see Fig. 152. By assumption, f_0 and f_1 are transverse to
L and, if we choose the homotopy to be stationary for the times $0 \leqslant t \leqslant 1/3$
and $2/3 \leqslant t \leqslant 1$, then it follows that $F | M \times (0, 1/3]$ and $F | M \times [2/3, 1)$ are
transverse to L. Using (14.7) we may replace the homotopy $F | M \times (0, 1)$
by a map which is transverse to L without altering it on the closed set
$M \times ([0, 1/3] \cup [2/3, 1])$. Consider now the pre-image $F^{-1}(L) \subset M \times [0, 1]$;
then $F^{-1}(L) \cap M \times (0, 1)$ is a submanifold of the same codimension as
L in N, and

$$F^{-1}(L) \cap [0, 1/3) = f_0^{-1}(L) \times [0, 1/3);$$

$$F^{-1}(L) \cap (2/3, 1] = f_1^{-1}(L) \times (2/3, 1].$$

Putting everything together, one sees that $F^{-1}(L)$ is a manifold with
boundary equal to $f_0^{-1}(L) + f_1^{-1}(L)$, hence *homotopic maps which are transverse to L have bordant pre-images.* The bordism class of $f^{-1}(L)$ is thus independent of which particular approximation to f, transverse to L, one takes.

Indeed, one only needs to assume that the original map f is continuous,
since any continuous map can be approximated by a differentiable one.

(14.8) **Theorem.** *Let $f: M \to N$ be a continuous map which is
differentiable on an open neighbourhood U of the closed set $A \subset M$. Then,
arbitrarily close to f, there exists a differentiable map $g: M \to N$, so that
$g | A = f | A$.*

Proof. Choose a closed embedding $N \subset \mathbb{R}^n$, and a tubular neighbourhood V on N in $\mathbb{R}^n$ with projection $\pi: V \to N$ (see (7.10), (12.11)). Now let
W be a neighbourhood of the graph of f in $M \times N$, so that

$$Q:= \{(p, q) \in M \times V \mid \pi(q) \in W\}$$

is a neighbourhood of the graph of f in $M \times V$. If the graph of a differentiable map $g: M \to \mathbb{R}^n$ lies in Q, then the graph of $\pi \circ g$ lies in W, so that we may assume that $N = \mathbb{R}^n$.

In this case we consider an ϵ-neighbourhood of f, where $\epsilon: M \to \mathbb{R}$ is a strictly positive function; choose with this a covering $\{U_\nu \mid \nu \in \mathbb{Z}\}$ of M with a subordinate differentiable partition of unity $\{\phi_\nu\}$ together with constants f_ν, so that $|f(p) - f_\nu| < \epsilon(p)$ for all $p \in U_\nu$, and so that $U_\nu \subset U$ for $\nu < 0$, and $U_\nu \subset M - A$ for $\nu \geqslant 0$. Then one sets

$$g(p) = \sum_{\nu < 0} f(p) \cdot \phi_\nu(p) + \sum_{\nu \geqslant 0} f_\nu \phi_\nu(p). \qquad \square$$

(14.9) Exercises

1 Let A_0, A_1, be disjoint closed subsets of the differentiable manifold M. Show that there exists a differentiable function $\alpha: M \to \mathbb{R}$ such that $0 \leqslant \alpha \leqslant 1$, $\alpha^{-1}\{0\} = A_0$, $\alpha^{-1}\{0\} = A_1$.

2 Let M be a compact connected differentiable manifold, and $A \subset M$ a non-empty closed subset. Show that there exists a vector field on M, which vanishes on A and only on A.
Hint: first construct a vector field for which the set of zeros is finite.

3 In the transversality theorem (14.6) we have assumed that $N \subset E$ is a submanifold. Show that the same theorem holds if one replaces this inclusion by an arbitrary differentiable map $g: N \to E$. In this case, one must formulate the transversality condition (on $s: M \to E$) as follows: if $p \in M$, $q \in N$ and $s(p) = g(q) = x \in E$, then $T_p s(T_p M) + T_q g(T_q N) = T_x E$.

4 Formulate and prove a generalisation of the transversality theorem for maps (14.7), which corresponds to the generalisation of (14.6) in exercise 3.

5 Let B be a manifold with boundary and let $L \subset M$ be a differentiable submanifold. Show that each continuous map $f: B \to M$ is homotopic to a map $g: B \to M$, so that $g^{-1}L \subset B$ is a differentiable submanifold with boundary and $\partial(g^{-1}L) = g^{-1}L \cap \partial B$.

6 Let M be an oriented differentiable manifold, and let $f_\nu: N_\nu \to M$, $\nu = 1, 2$, be differentiable maps of oriented closed manifolds of complementary dimensions, that is, $\dim(N_1) + \dim(N_2) = \dim(M)$. The *intersection number* $[f_1] \cdot [f_2] \in \mathbb{Z}$ is then defined as follows: choose a map g homotopic to f_1 and transverse in the sense of exercise 3 to f_2. Then the fibre product $F := \{(p, q) \in N_1 \times N_2 \mid g(p) = f_2(q)\}$ is finite (5.14, 11) and, for each pair $(p, q) \in F$, one has an isomorphism of oriented vector spaces $T_p(N_1) \oplus T_q(N_2) \xrightarrow{(Tg, Tf_2)} T_{g(p)}M$, and we may set $\epsilon(p, q) = \pm 1$, depending on whether this isomorphism preserves or reverses orientatation. Then

$$[f_1] \cdot [f_2] := \sum_F \epsilon(p, q), \quad \text{see Fig. 153.}$$

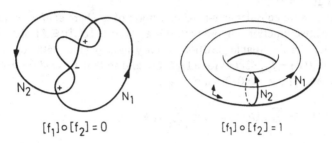

$$[f_1] \circ [f_2] = 0 \qquad\qquad [f_1] \circ [f_2] = 1$$

Fig. 153

Show that the intersection number is well defined and depends only on the homotopy classes of the maps f_ν. Furthermore, $[f_1] \cdot [f_2] = (-1)^{n_1 \cdot n_2}[f_2] \cdot [f_1]$, $n_\nu := \dim N_\nu$.

7 For a connected manifold M let $\pi_n M$ be the set of homotopy classes of continuous maps $S^n \to M$. Show that if $n < k$, then $\pi_n S^k = 0$. *Hint*: (14.8), (6.1).

8 Let $i: \{p\} \to S^n$ be the inclusion of a point. Show that the following map $s: \pi_n S^n \to \mathbb{Z}$, $n > 0$ is surjective: if $\alpha \in \pi_n S^n$ is represented by the map $a: S^n \to S^n$, then $s(\alpha) := [a] \cdot [i]$, see exercises 6 and 7.

9 More generally than in exercise 8 let M be a closed, oriented, connected, differentiable n-dimensional manifold and Π the set of homotopy classes of continuous maps $f: M \to S^n$. If $i: \{p\} \to S^n$ is again the inclusion of a point, then $f \mapsto [f] \cdot [i]$ is a surjection $\Pi \to \mathbb{Z}$.

10 Show that the map $\Pi \to \mathbb{Z}$ in exercise 9 is bijective hence, in particular, that $\pi_n S^n \cong \mathbb{Z}$. *Hint*: use (10.3).

11 Let $s: M \to TM$ be a vector field on a closed oriented manifold (*TM* possesses a canonical orientation (4.22, 5), (11.7, 2)). The number $\chi(M) := [s] \cdot [s]$ is called the Euler characteristic of M. Show that $\chi(M)$ depends only on M (exercise 6). If there exists a nowhere vanishing vector field on M, then $\chi(M) = 0$.

12 Show that $\chi(S^{2n+1}) = 0$, $\chi(S^{2n}) = 2$ (see exercise 11). *Hint*: S^{2n+1} is the unit sphere in $\mathbb{C}^{n+1}$, and one can construct a nowhere vanishing vector field. For S^{2n} consider the vector field induced by rotation about an axis.

References

1. Godement, R. (1964). *Topologie algébrique et théorie des faisceaux*. Paris: Hermann.
2. Lang, S. (1969). *Analysis I*. London: Addison-Wesley.
3. Lang, S. (1972). *Differential manifolds*. Reading, Mass.: Addison-Wesley.
4. Milnor, J. (1958). Differential topology (mimeograph) Princeton.
5. Milnor, J. (1961). Differentiable structures (mimeograph) Princeton.
6. Milnor, J. (1965). *Topology from the differentiable viewpoint*. Charlottesville, The University Press of Virginia.
7. Narasimhan, R. (1968). *Analysis on real and complex manifolds*. Amsterdam: North-Holland.
8. Kelley, J. H. (1968). *General topology*. Princeton, New Jersey: Van-Nostrand.
9. Sternberg, S. (1964). *Lectures on differential geometry*. New Jersey: Prentice Hall Inc.

Index of symbols

M^n	n-dimensional manifold 1	f^*E	induced bundle 27
		$\oplus$	Whitney sum 33
(h, U)	chart 2	$\otimes$	tensor product 34
$\|x\|$	Euclidean valuation 4	E^*	dual bundle 34
$[x]$	equivalence class, in particular homogenous coordinates 4	Λ^k	exterior power 34
		$\mathfrak{o}, \mathfrak{o}_x$	orientation 34
		$\tilde{X}(E)$	orientation cover 35
$[0, 1], [0, 1),$ $(0, 1)$	unit interval (closed, half-open, open) 7	$\perp X$	normal bundle 38
$+$	differentiable sum 8	$\bar{\{\}}, \bar{W}$	closure 41
$\times$	differentiable (Cartesian) product 8	Δ_M	diagonal 42, 121
		$\|f\|_K$	seminorm on $C^\infty(M, N)$ 65
f^*	induced map 12, 14		
$\langle \, , \rangle$	inner or scalar product 13	Φ, Φ_t	local flow 74
		$\dot{\alpha}$	velocity vector (curve, field) 76, 79
$\sim$	equivalence relation 14, 103	$\#$	connected sum 98
$\bar{f}: (M, p)$ $\to (N, q)(\mathscr{E})$	germ(s) 14	$\cup_\alpha$	identification by means of α 103
		$\ddot{\gamma}$	109
$\Box$	conclusion of proof 15	∂	boundary of a manifold 131
$[E, \pi, X]$	vector bundle 22		
E_x	fibre 22	M/τ	quotient manifold 135
f_x	fibre mapping 22	$[f_1] \circ [f_2]$	intersection number 151
$E^\perp$	orthogonal complement 24		
$E\|X$	restriction of a bundle 25		

155

Subject Index

157